Bending and Torsion of thin walled Beams with variable, open Cross Sections

von

Dr. sc. techn. Suresh Lonkar

Institut für Baustatik

Eidgenössische Technische Hochschule Zürich

Zürich

November 1968

ISBN 978-3-0348-4069-9   ISBN 978-3-0348-4143-6 (eBook)
DOI 10.1007/978-3-0348-4143-6

F O R E W O R D

The theory of thin walled beams with open cross section has
many applications in practice, e.g. in the analysis of girder bridges. For beams of constant section solutions have
been available in the technical literature. However, few
publications exist which deal with beams of variable cross
section. In this study a solution for beams with unsymmetrical and variable cross section is presented. A corresponding
computer program for pratical application has been developed.

The author has prepared this study in partial fulfilment of
his doctoral work at the Institute of Structural Engineering,
Department of Civil Engineering. During his stay he received
a scholarship of the Department of the Interior of the Swiss
Federal Government.

Swiss Federal Institute of Technology

Zurich, January 1969

Prof. Dr. B. Thürlimann

BENDING AND TORSION OF THIN WALLED STRAIGHT AND CURVED

BEAMS WITH VARIABLE OPEN CROSS SECTION

CONTENTS

1. INTRODUCTION

1.1 Description of the Problem

The theory outlined in the present dissertation aims at the
general solution of the problem of bending and torsion of
thin walled beams with open cross section. The longitudinal
axis of the beam may be straight or curved. The conditions
of support of the beam are arbitrary. The cross section may
be unsymmetrical and variable along the axis.

1.2 Literature Review

The general problem of torsion of thin walled beams with
constant open cross section has been trated by many authors.
A comprehensive review of the pertinent literature on the
subject of warping torsion may be found in the last chapter
of the well-known book by V.Z. Vlassov [1].

On the problem of torsion of thin walled beams with variable
open section work has been done by Z. Cywinski [2], G. Becker
[3] and Bažant [4]. Cywinski obtains the differential equations of torsion for such beams from the minimum principle
of potential energy. He uses the finite difference approach
for the solution of the differential equations with variable
coefficients. He compares the results of his theoretical solution with those obtained from tests on a plexi-glass model.

Becker considers the equation of torsion of a curved beam of
constant monosymmetrical section, as derived by Vlassov. For
variable sections he imagines the beam to be made up of short
pieces of constant section and uses the solution of the sixth
order differential equation mentioned earlier. A similar
approach for the analysis of box beams of variable section

has been suggested by Heilig [5]

Both the papers [2] and [3] refer to the case of cross sections with one axis of symmetry. The present work treats the general case of an unsymmetrical section. In paragraphs 5.1 and 5.2 comparisons are made between the solutions of Cywinski and Becker with those of the present theory.

E. Karamuk [6] considers the solution of the differential equation of torsion of a beam of monosymmetrical open section in two parts. In his approach first the St. Venant Torsional Rigidity is neglected and only the warping Rigidity is considered. The differential equation is then similar to that of plane bending of a beam. Next only the St. Venant Rigidity is considered and the second order homogeneous equation is solved. At discrete points the two solutions are coupled to fulfill the conditions of compatibity. The redundant forces are then calculated as in the force method. The Folded Plate Theory is also discussed.

2. DIFFERENTIAL EQUATIONS AND SOLUTIONS

2.1 Differential Equations of Equilibrium of a Thin

Walled Beam of Constant Open Cross Section

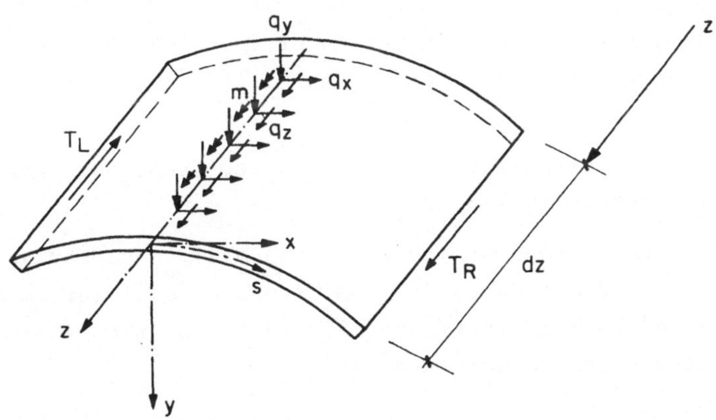

FIG. 1

A complete derivation of the differential equation using
the arbitrary coordinate system may be found in reference
[1].

Fig. 1 shows an element of a thin walled beam of open cross
section. The external loads q_x, q_y, and q_z act along the
positive directions of the x, y, z axes respectively, when
looking from the positive z side. The external torsional moment m acts clockwise and the positive direction of the moment vector coincides with the positive direction of the z-
axis, consistent with the left handed system of axes x, y,
z. It may be noted that these are chosen arbitrarily. In the
plane of cross section, the x and y axes are neither the
axes of symmetry nor the principal axes. The s axis is along

the centre line of the walls of the members and the positive s direction is in the clockwise sense as seen from the positive z side.

The external loads q_x, q_y, and q_z possess the unit kg/cm and the moment m kg.cm/cm. T_L and T_R are the external shear forces acting on the left and right hand edges $S = S_L$ and $S = S_R$ respectively.

Next we consider the deformation of the element under the action of the external loads. The following assumptions are made:

1) The material of the beam is elastic. The Stress-Strain relationship is linear.

2) The deformations of the beam are small in comparison with its dimensions. The conditions of equilibrium of the undeformed system are also valid for the deformed system (first order theory).

3) The dimensions of the beam are consistent with the description "thin walled". The normal and shear stresses are constant over the thickness of each element of the cross section.

4) The cross sectional shape remains unchanged. Hence, under external loads, each point in the cross section undergoes rigid body displacements in the plane of cross section. These displacements correspond to three degrees of freedom, and can be expressed as functions of:

   a) The translations U_x and U_y of the origin of coordinates O, and

   b) The rotation Θ of the cross section with respect to the system of axes x, y, z. This is shown in Fig. 2.

5) During deformation the shear strain of the middle surface is relatively small and hence neglected. Thus a normal to the section remains normal. The warping of the

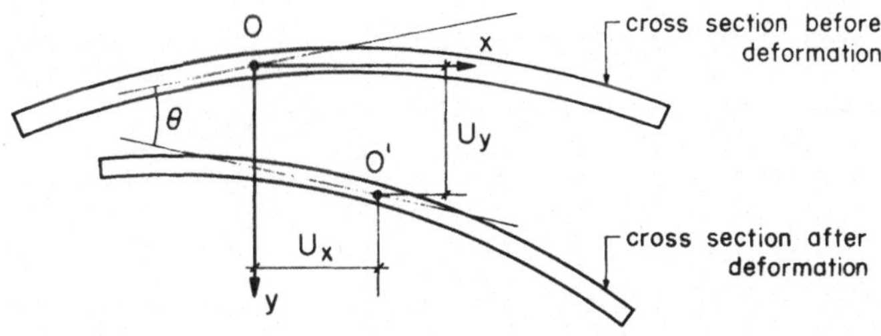

FIG. 2

cross section in the direction of the beam axis in determined by this condition.

A point P in the cross section has coordinates x, y, z, w. The coordinate w with the dimensions of an area is defined as follows: See Fig. 3.

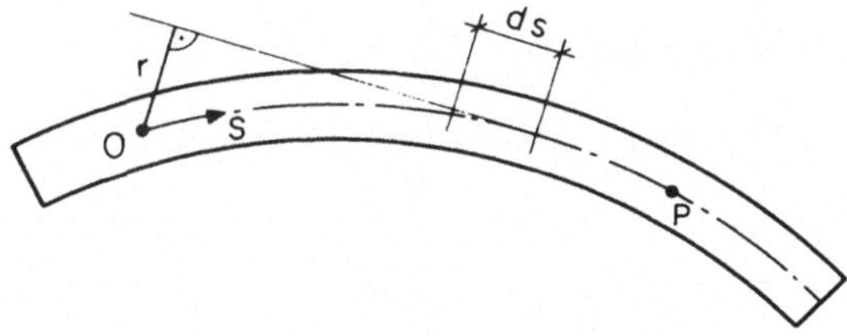

FIG. 3

ds is an elemental arc length along the s axis between origin O and the point P and r the distance between O and the tangent to ds. Then

$$w = \int_{O}^{P} r\, ds$$

The point P has three displacements Ux_p, Uy_p, and Uz_p. These can be expressed in terms of the displacements of the coordinate origin O as follows:

$$U_{x_p} = U_x - y \cdot \theta \qquad \text{——} \ (1)$$

$$U_{y_p} = U_y + x \cdot \theta \qquad \text{——} \ (2)$$

$$U_{z_p} = U_z - \frac{dU_x}{dz} \cdot x - \frac{dU_y}{dz} \cdot y - \frac{d\theta}{dz} \cdot w \qquad \text{——} \ (3)$$

In equation (3), the first three terms on the right hand side represent plane bending and the last term indicates the warping of the cross section. This equation can be derived from the condition of zero shear strain of the middle surface of the beam, and is given here without proof.

In many practical cases, the longitudinal edges of the beam are free, and hence the edge shears T_L and T_R vanish. The differential equations of equilibrium may be written as follows:

$$L \cdot U_z{}'' - \left(\frac{Sy}{F}\right) U_x{}''' - \left(\frac{Sx}{F}\right) U_y{}''' - \left(\frac{Sw}{F}\right) \theta{}''' = -\frac{q_z L^3}{EF} \qquad \text{——} \ (4)$$

$$-L \cdot \left(\frac{Sy}{Iy}\right) U_z{}''' + U_x{}^{\overline{IV}} + \left(\frac{Ixy}{Iy}\right) U_y{}^{\overline{IV}} + \left(\frac{Iwx}{Iy}\right) \theta{}^{\overline{IV}} = \frac{q_x L^4}{EIy} \qquad \text{——} \ (5)$$

$$-L \cdot \left(\frac{Sx}{Ix}\right) U_z{}''' + \left(\frac{Ixy}{Ix}\right) U_x{}^{\overline{IV}} + U_y{}^{\overline{IV}} + \left(\frac{Iwy}{Ix}\right) \theta{}^{\overline{IV}} = \frac{q_y L^4}{EIx} \qquad \text{——} \ (6)$$

$$-L \cdot \left(\frac{Sw}{Iw}\right) U_z{}''' + \left(\frac{Iwx}{Iw}\right) U_x{}^{\overline{IV}} + \left(\frac{Iwy}{Iw}\right) U_y{}^{\overline{IV}} + \theta{}^{\overline{IV}} - \frac{GK}{EIw} \cdot L^2 \cdot \theta{}'' = \frac{m L^4}{EIw}$$

$$\text{——} \ (7)$$

Equations (4) to (7) are due to Vlassov [1]. They are expressed here in a slightly different form for convenience. It may be particularly noted that the symbol ' represents the derivative with respect to ζ, whereby $\zeta = \frac{z}{L}$ is a dimensionless parameter, L being a reference length of the beam.

The various symbols used in this paragraph and in all further equations are explained in chapter 7.

2.2 Homogeneous Solutions of the System of Differential

Equations

Vlassov [1] has given a solution for a special coordinate system using principle axes reduced to the shear center. Here a general solution for the arbitrary coordinate system is presented.

Differentiating equation (4) with respect to ζ, the term $L \cdot U_z'''$ is expressed as:

$$L\, U_z''' = \left(\frac{Sy}{F}\right) Ux^{\overline{IV}} + \left(\frac{Sx}{F}\right) Uy^{\overline{IV}} + \left(\frac{Sw}{F}\right) \theta^{\overline{IV}} - \left(\frac{L^3}{EF}\right) qz' \qquad \text{---} \ (8)$$

Substituting for $L \cdot U_z'''$ in equations (5), (6), and (7), these can be rewritten in the following form:

$$a_{11}\, Ux^{\overline{IV}} + a_{12}\, Uy^{\overline{IV}} + a_{13}\, \theta^{\overline{IV}} = X_2 \cdot qx - X_1 \cdot g_9 \cdot qz' \qquad \text{---} \ (9)$$

$$a_{21}\, Ux^{\overline{IV}} + a_{22}\, Uy^{\overline{IV}} + a_{23}\, \theta^{\overline{IV}} = X_3 \cdot qy - X_1 \cdot g_6 \cdot qz' \qquad \text{---} \ (10)$$

$$a_{31}\, Ux^{\overline{IV}} + a_{32}\, Uy^{\overline{IV}} + a_{33}\, \theta^{\overline{IV}} - a_{34}\, \theta'' = X_4 \cdot m - X_1 \cdot g_{12} \cdot qz' \qquad \text{---} \ (11)$$

From equation (10)

$$Uy^{\overline{IV}} = -\frac{a_{21}}{a_{22}} Ux^{\overline{IV}} - \frac{a_{23}}{a_{22}} \theta^{\overline{IV}} + \frac{1}{a_{22}} X \cdot qy - \frac{g_6}{a_{22}} \cdot X_2\, qz' \qquad \text{---} \ (12)$$

This expression for $U_y^{\overline{IV}}$ is substituted in equations (9) and (11):

$$b_{11}\, Ux^{\overline{IV}} + b_{12}\, \theta^{\overline{IV}} = X_2 \cdot qx + X_3 \cdot b_{14}\, qy - X_1 \cdot g_{15} \cdot qz' \qquad \text{---} \ (13)$$

$$b_{21}\, Ux^{\overline{IV}} + b_{22}\, \theta^{\overline{IV}} - b_{23}\, \theta'' = X_4 \cdot m + X_3 \cdot b_{24}\, qy - X_1 \cdot g_{16} \cdot qz' \qquad \text{---} \ (14)$$

From equation (13)

$$U_x^{\overline{IV}} = -\frac{b_{12}}{b_{11}}\,\theta^{\overline{IV}} + \frac{x_2}{b_{11}}\,q_x + x_3\cdot\frac{b_{14}}{b_{11}}\,q_y - \frac{g_{15}}{b_{11}}\,x_1\cdot q_z' \qquad\text{---- (15)}$$

This expression for $U_x^{\overline{IV}}$ is substituted in equation (14), which yields after rearrangement:

$$C_{11}\,\theta^{\overline{IV}} - C_{12}\,\theta'' = x_4\cdot m + \gamma\cdot x_2\cdot q_x + g_{17}\cdot x_3\cdot q_y - g_{18}\cdot x_1\cdot q_z'$$

$$\text{---- (16)}$$

Equation (16) is rewritten in the following form:

$$\theta^{\overline{IV}} - \beta^2\,\theta'' = A\cdot m + B\cdot q_x + C\cdot q_y - D\cdot q_z' \qquad\text{---- (17)}$$

This is the final differential equation for the problem of bending and torsion of a thin walled beam with constant open section.

The homogeneous solution for θ in equation (17) is of the form:

$$\theta_h = C_1 + C_2\,\zeta + C_3\,\sinh\beta\zeta + C_4\,\cosh\beta\zeta \qquad\text{---- (18)}$$

By inspecting (15), the homogeneous solution for U_x is obtained as

$$U_{xh} = a\,\theta_h + C_5 + C_6\,\zeta + C_7\,\zeta^2 + C_8\,\zeta^3 \qquad\text{---- (19)}$$

Equation (12) is solved for U_y as:

$$U_{yh} = g_1\cdot\theta_h + g_2\,U_{xh} + C_9 + C_{10}\,\zeta + C_{11}\,\zeta^2 + C_{12}\,\zeta^3 \qquad\text{---- (20)}$$

Finally, integration of equation (4) twice with respect to ζ yields:

$$U_{zh} = \frac{1}{L}\left\{(g_3\,\theta_h' + g_4\,U_{xh}' + g_5\,U_{yh}') + C_{13} + C_{14}\,\zeta\right\} \qquad\text{---- (21)}$$

The solutions (18) to (21) involve 14 constants of integration C_1 to C_{14}. They are calculated from the boundary conditions.

2.3 Particular Integrals

The particular integrals for the differential equations (17), (15), (12), and (4) can be obtained straight forward when the loadings q_x, q_y, q_z, and m are constant or known explicitly as continuous functions of the variable ζ. '

In the following, the particular solutions for Θ, U_x, U_y, and U_z are given for constant m, q_x, q_y, and q_z being a linear function of ζ.

In this case, the particular integral for Θ in equation (17) is obtained by inspection as:

$$\theta_p = -\frac{A}{2\beta^2}\zeta^2 \cdot m - \frac{B}{2\beta^2}\zeta^2 \, qx - \frac{C}{2\beta^2}\, qy + \frac{D}{2\beta^2}\, qz' \qquad \text{—— (22)}$$

or

$$\theta_p = A_0 \cdot m + B_0 \cdot qx + C_0 \cdot qy - D_0 \cdot qz' \qquad \text{—— (23)}$$

In the same way equation (15) yields:

$$Ux_p = \alpha \cdot \theta_p + a_2 \cdot \zeta^4 \cdot qx + a_3 \cdot \zeta^4 \cdot qy - a_4 \cdot \zeta^4 \cdot qz' \qquad \text{—— (24)}$$

Going back to equation (12), the particular solution for U_y is seen to be:

$$Uy_p = g_1 \cdot \theta_p + g_2 \cdot Ux_p + r_1 \cdot \zeta^4 \cdot qy - r_2 \cdot \zeta^4 \cdot qz' \qquad \text{—— (25)}$$

Finally from equation (4) the particular solution for U_z is obtained as:

$$Uz_p = \frac{1}{L}\left\{ (g_3 \cdot \theta_p' + g_4 \cdot Ux_p' + g_5 \cdot Uy_p') - r_3 \cdot \zeta^2 \cdot (k_1 \zeta + 3k_2) \right\}$$
$$\text{—— (26)}$$

The definition of the various symbols is given in chapter 7.

The complete solutions for the differential equations (4) to (7), namely for the displacements U_x, U_y, U_z of the co-ordinate origin O, and the angle of rotation θ of the cross section are thus as follows:

$$\theta = \theta_h + \theta_p \qquad \text{— Equations (18) and (23)}$$

$$U_x = U_{xh} + U_{xp} \qquad \text{— Equations (19) and (24)}$$

$$U_y = U_{yh} + U_{yp} \qquad \text{— Equations (20) and (25)}$$

$$U_z = U_{zh} + U_{zp} \qquad \text{— Equations (21) and (26)}$$

2.4 Remarks on the Form of the Differential Equations

Referring to equations (4) to (7), the following points be noted:

Equation (4) expresses the equilibrium of the internal and external forces along the z or longitudinal axis. Equation (5) refers to bending in the xz plane. The bending of the element in the yz plane is expressed by equation (6). The equilibrium of the external and internal torsional moments is contained in differential equation (7).

In the case of a beam with two axes of symmetry, if x and y are the principal axes, equations (4), (5), (6), and (7) contain terms only in U_z, U_x, U_y, and Θ respectively. The coefficients of the other terms, for example of U_x, U_y, and Θ in equation (4), would vanish. The solutions for U_x, U_y, U_z, and Θ will then be independant of eachother. In a particular problem this would be automatically taken care of in solutions (18) to (21) and (23) to (26).

Choosing the x, y, and z axes arbitrarily and then getting the solutions for the displacements of the arbitrary origin O, namely U_x, U_y, U_z and the rotation Θ of the cross section leads to a solution for the general problem of bending and torsion of a thin walled beam of unsymmetrical open cross section.

2.5 Expressions for the Sectional Forces

In paragraphs 2.2 and 2.3 solutions were obtained for U_x, U_y, U_z, and Θ, which satisfy the differential equations (4) to (7). The internal forces acting on a cross section can now be expressed as functions of the derivates of these displacements.

Referring to Fig. 4 the basic expressions for the normal and shear stresses at a point P with coordinates x, y, z, s, w are as follows:

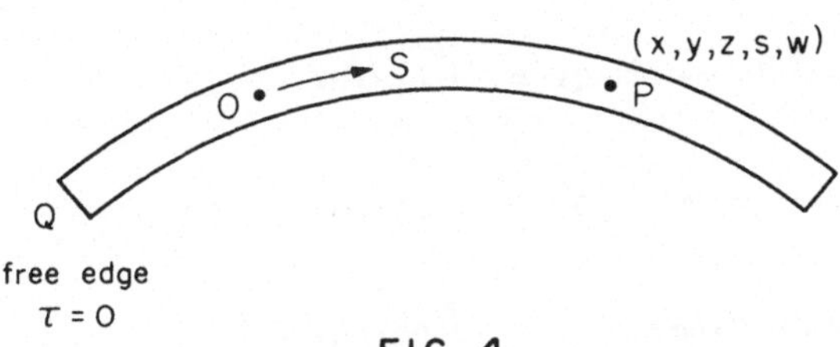

FIG. 4

Normal Stress:

$$\sigma = \frac{E}{L^2}(L \cdot U_z' - U_x'' \cdot x - U_y'' \cdot y - \theta'' \cdot w) \quad \text{—— (27)}$$

Shear Stress:

$$\tau = \frac{E}{L^2}\left(-\frac{F(s)}{t} \cdot L \cdot U_z'' + \frac{Sy(s)}{t} \cdot U_x''' + \frac{Sx(s)}{t} \cdot U_y''' + \frac{Sw(s)}{t} \cdot \theta'''\right)$$

$$\text{—— (28)}$$

The quantities $F(s)$, $S_y(s)$, $S_x(s)$, and $S_w(s)$ are defined as follows:

$$F(s) = \int_Q^P dF \qquad\qquad Sy(s) = \int_Q^P y\,dF$$

$$Sx(s) = \int_Q^P x\,dF \qquad\qquad Sw(s) = \int_Q^P w\,dF$$

The point Q lies on the free edge where the shear stress vanishes. While going from Q to P along the centre line of the wall the movement is in the positive s sense. i.e. clockwise.

The sectional forces are defined as follows:

1) Normal force $\quad N = \int_F \sigma\,dF$

2) Bending Moment $\quad My = -\int_F \sigma x\,dF$

3) Bending Moment $\quad Mx = \int_F \sigma y\,dF$

4) Warping Moment $\quad Mw = \int_F \sigma w\,dF$

5) Shear Force $\quad Qx = \int_S (\tau t)\,dx$

6) Shear Force $\quad Qy = \int_S (\tau t)\,dy$

7) Torsional Moment $\quad T = \int_W (\tau t)\,dw + \frac{GK}{L} \cdot \theta'$

Where Warping Torsional Moment $\quad Ts = \dfrac{GK}{L} \cdot \theta'$

St. Venant Torsional Moment $\quad Tw = \displaystyle\int_w (\tau\,t)\,dw$

On using equations (27) and (28) and carrying out the integrations, the following expressions are obtained for the Section Forces:

$$N = \frac{EF}{L^2}\left(L \cdot Uz' - g_4 \cdot Ux'' - g_5 \cdot Uy'' - g_3\,\theta''\right) \quad\text{---- (29)}$$

$$Mx = \frac{EIx}{L^2}\left(L \cdot g_6\ Uz' - g_7 \cdot Ux'' - Uy'' - g_8 \cdot \theta''\right) \quad\text{---- (30)}$$

$$My = -\frac{EIy}{L^2}\left(L \cdot g_9 \cdot Uz' - Ux'' - g_{10}\ Uy'' - g_{11}\,\theta''\right) \quad\text{---- (31)}$$

$$Mw = \frac{EIw}{L^2}\left(L \cdot g_{12} \cdot Uz' - g_{13} \cdot Ux'' - g_{14}\ Uy'' - \theta''\right) \quad\text{---- (32)}$$

$$Ts = \frac{GK}{L}\,\theta' \quad\text{---- (33a)}$$

$$Tw = \frac{EIw}{L^3}\left(g_{12} \cdot L \cdot Uz'' - g_{13} \cdot Ux''' - g_{14}\ Uy''' - \theta'''\right) \quad\text{---- (33b)}$$

$$T = Ts + Tw = \frac{EIw}{L^3}\left(g_{12} \cdot L \cdot Uz'' - g_{13} \cdot Ux''' - g_{14} \cdot Uy''' - \theta''' + \phi_1 \cdot \theta'\right) \quad\text{---- (33)}$$

$$Qx = \frac{EIy}{L^3}\left(L \cdot g_9 \cdot Uz'' - Ux''' - g_{10}\ Uy''' - g_{11}\,\theta'''\right) \quad\text{---- (34)}$$

$$Qy = \frac{EIx}{L^3}\left(L \cdot g_6\ Uz'' - g_7 \cdot Ux''' - Uy''' - g_8\ \theta'''\right) \quad\text{---- (35)}$$

The various symbols used are explained in chapter 7. In the appendix a matrix representation of relations (29) to (35) is given.

3. THE FINITE ELEMENTS METHOD FOR BEAMS OF VARIABLE

SECTION

3.1 The Transport Matrix Method

This method is discussed for Beam Bending problems by
Zurmühl [7]. Still, the essentials will be presented here
for the sake of clarity with the help of a simple illustrative example. In the following simple bending of a cantilever beam is considered.

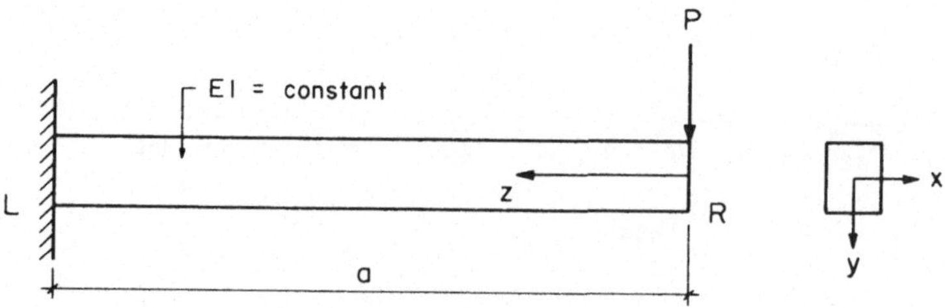

FIG. 5

Fig. 5 shows a cantilever beam LR of length a. The beam is
prismatic and its cross section has two axes of symmetry x,
and y. It carries a concentrated load P at the free end R.
Using the transport matrix method the bending moment and
the shear force at the fixed end L and the slope and vertical deflection at the free end R are to be calculated.

The Differential Equation of equilibrium in the plane yz of
the load is:

$$\frac{d^4 \cdot U_y}{d \cdot \zeta^4} = \frac{q_y \cdot a^4}{E\,I} = 0 \quad , \qquad \zeta = \frac{z}{a}$$

The solution for U_y is

$$U_y = A + B\zeta + C\zeta^2 + D\zeta^3 \, , \qquad \text{--- (36)}$$

A, B, C, D being the constants of integration

Slope
$$\frac{dU_y}{dz} = \frac{1}{a} \cdot \frac{dU_y}{d\zeta} = \frac{1}{a}\left[B + C \cdot 2\zeta + D \cdot 3\zeta^2\right] \qquad \text{--- (37)}$$

Bending Moment $M = -EI\dfrac{d^2U_y}{dz^2} = -\dfrac{EI}{a^2} \cdot \dfrac{d^2U_y}{d\zeta^2} = -\dfrac{EI}{a^2}\left[2C + 6D\zeta\right]$
$$\text{--- (38)}$$

Shear Force $Q = \dfrac{dM}{dz} = \dfrac{1}{a} \cdot \dfrac{dM}{d\zeta} = -\dfrac{EI}{a^3} \cdot 6D \qquad \text{--- (39)}$

Equations (36) to (39) are written in Matrix form as follows

$$
\begin{bmatrix} U_y \\[2mm] \dfrac{1}{a}\dfrac{dU_y}{d\zeta} \\[2mm] M \\[2mm] Q \end{bmatrix}
=
\begin{bmatrix}
1 & \zeta & \zeta^2 & \zeta^3 \\[2mm]
0 & \dfrac{1}{a} & \dfrac{2\zeta}{a} & \dfrac{3\zeta^2}{a} \\[2mm]
0 & 0 & -\dfrac{2EI}{a^2} & -\dfrac{6EI}{a^2}\zeta \\[2mm]
0 & 0 & 0 & -\dfrac{6EI}{a^3}
\end{bmatrix}
\begin{bmatrix} A \\[2mm] B \\[2mm] C \\[2mm] D \end{bmatrix}
\qquad \text{--- (40)}
$$

<u>Boundary Conditions</u>

a) Free End : $\zeta = 0$,   1) $M = 0$, 2) $Q = -P$

b) Fixed End : $\zeta = 1$,   3) $U_y = 0$, 4) $\dfrac{1}{a}\dfrac{dU_y}{d\zeta} = 0$

For a particular value of , the quantities on the left hand side of equation (40) constitute a vector called the "State Vector".

The State Vector $\begin{bmatrix} S_R \end{bmatrix}$ at the free end is given by:

$$
[S_R] = \begin{bmatrix} Uy_R \\[4pt] \dfrac{1}{a}\left(\dfrac{dUy}{d\zeta}\right)_R \\[4pt] 0 \\[4pt] -P \end{bmatrix} = \begin{bmatrix} 1 & 0 & 0 & 0 \\[4pt] 0 & \dfrac{1}{a} & 0 & 0 \\[4pt] 0 & 0 & -\dfrac{2EI}{a^2} & 0 \\[4pt] 0 & 0 & 0 & -\dfrac{6EI}{a^3} \end{bmatrix} \begin{bmatrix} A \\[4pt] B \\[4pt] C \\[4pt] D \end{bmatrix}
$$

$$\text{—— (41)}$$

In Matrix notation equation (41) is written as

$$[S_R] \;=\; [X]\cdot[CON] \qquad \text{—— (42)}$$

Here $[X]$ is the 4 x 4 Matrix on the right hand side of equation (41), and $[CON]$ is the vector of the four constants of integration A, B, C, D.

Similarly, the State Vector at the fixed end L is given by:

$$
[S_L] = \begin{bmatrix} 0 \\[4pt] 0 \\[4pt] M_L \\[4pt] Q_L \end{bmatrix} = \begin{bmatrix} 1 & 1 & 1 & 1 \\[4pt] 0 & \dfrac{1}{a} & \dfrac{2}{a} & \dfrac{3}{a} \\[4pt] 0 & 0 & -\dfrac{2EI}{a^2} & -\dfrac{6EI}{a^2} \\[4pt] 0 & 0 & 0 & -\dfrac{6EI}{a^3} \end{bmatrix} \begin{bmatrix} A \\[4pt] B \\[4pt] C \\[4pt] D \end{bmatrix}
$$

Or in Matrix notation:

$$[S_L] \;=\; [Y]\cdot[CON] \qquad \text{—— (43)}$$

Premultiplying both sides of equation (42) by the inverse matrix $[X]^{-1}$, it is rewritten as

$$[CON] = [X]^{-1} \cdot [S_R] \qquad \text{—— (44)}$$

Substituting this expression for the vector $[CON]$ in equation (43):

$$[S_L] = [Y] \cdot [X]^{-1} \cdot [S_R]$$
$$[TR]_R^L \cdot [S_R] \qquad \text{—— (45)}$$

In equation (45) the "Transport Matrix" $[T_R]_R^L$ is given by:

$$[TR]_R^L = [Y] \cdot [X]^{-1}$$

X being a diagonal matrix,

$$\text{Inverse Matrix} \quad [X]^{-1} = \begin{bmatrix} 1 & 0 & 0 & 0 \\ 0 & a & 0 & 0 \\ 0 & 0 & -\dfrac{a^2}{2EI} & 0 \\ 0 & 0 & 0 & -\dfrac{a^3}{6EI} \end{bmatrix} \qquad \text{—— (46)}$$

By the rules of Matrix multiplication, the Transport Matrix $[TR]_R^L$ is obtained as:

$$[TR]_R^L = \begin{bmatrix} 1 & a & -\dfrac{a^2}{2EI} & -\dfrac{a^3}{6EI} \\ 0 & 1 & -\dfrac{a}{EI} & -\dfrac{a^2}{2EI} \\ 0 & 0 & 1 & a \\ 0 & 0 & 0 & 1 \end{bmatrix} \qquad \text{—— (47)}$$

Substituting this expression for $[TR]_R^L$, Matrix equation (45) is rewritten as:

$$\begin{bmatrix} 0 \\ 0 \\ M_L \\ Q_L \end{bmatrix} = \begin{bmatrix} 1 & a & -\dfrac{a^2}{2EI} & -\dfrac{a^3}{6EI} \\ 0 & 1 & -\dfrac{a}{EI} & -\dfrac{a^2}{2EI} \\ 0 & 0 & 1 & a \\ 0 & 0 & 0 & 1 \end{bmatrix} \begin{bmatrix} U_{yR} \\ \dfrac{1}{a}\left(\dfrac{dU_y}{d\zeta}\right)_R \\ 0 \\ -P \end{bmatrix} \quad\text{----- (48)}$$

Equation (48) yields:

$$Q_L = -P$$

$$M_L = -Pa$$

$$(U_y)_R = \frac{Pa^3}{3EI}$$

$$\frac{1}{a}\left(\frac{dU_y}{d\zeta}\right)_R = -\frac{Pa^2}{2EI}$$

The results are consistent with the chosen sign convention.

This method is now extended to the problem of bending and torsion of thin walled beams. In this case the State Vector has fifteen rows. These are:

The seven Deformations:

$$U_x , \ U_x' , \ U_y , \ U_y' , \ U_z , \ \theta , \ \text{and} \ \theta'$$

The seven Section Forces:

$$N , \ M_x , \ M_y , \ Q_x , \ Q_y , \ T , \ \text{and} \ M_w$$

and the term Unity which takes care of the particular solutions and their derivatives in the matrix operations.

For the sake of convenience, the above quantities, namely the Deformations and Section Forces are multiplied by suitable multipliers to get well conditioned transport matrices.

There are fourteen boundary conditions in this case, seven at each of the two boundaries of the beam. They may be statical, geometrical, or mixed. Some of the usual conditions of support are:

a) <u>Fixed End:</u> Only geometrical conditions

$$1) \quad U_x \quad = \quad 0$$

$$2) \quad U_x' \quad = \quad 0$$

$$3) \quad U_y \quad = \quad 0$$

$$4) \quad U_y' \quad = \quad 0$$

$$5) \quad U_z \quad = \quad 0$$

$$6) \quad \theta \quad = \quad 0$$

$$7) \quad \theta' \quad = \quad 0$$

b) <u>Free End:</u> Only statical conditions

$$1) \quad N \quad = \quad 0$$

$$2) \quad M_x \quad = \quad 0$$

$$3) \quad M_y \quad = \quad 0$$

$$4) \quad M_w \quad = \quad 0$$

$$5) \quad Q_x \quad = \quad 0$$

$$6) \quad Q_y \quad = \quad 0$$

$$7) \quad T \quad = \quad 0$$

c) <u>Hinged End:</u> In this case the end of the beam is free to rotate in the yz plane but prevented from rotating in the xy plane.
There are three geometrical and four statical conditions.

$$
\begin{aligned}
1)\quad &\theta && = && 0 \\
2)\quad &U_x && = && 0 \\
3)\quad &U_y && = && 0 \\
4)\quad &N && = && 0 \\
5)\quad &M_x && = && 0 \\
6)\quad &M_y && = && 0 \\
7)\quad &M_w && = && 0
\end{aligned}
$$

The case of intermediate supports is discussed in paragraph 3.4.

3.2 Transport Matrix for Curved Beams

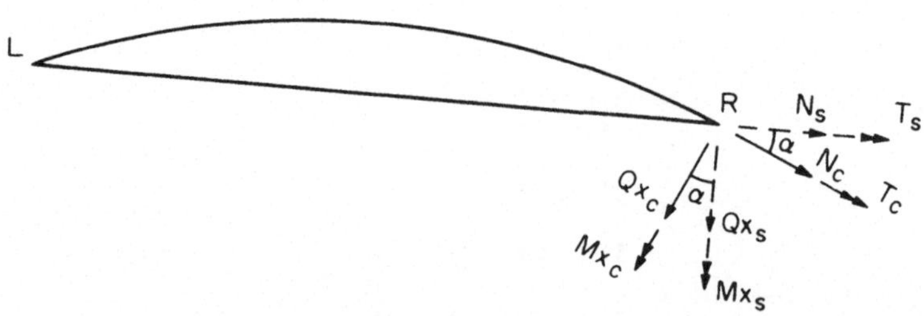

FIG. 6

Fig. 6 shows a curved Finite Element LR. If the difference between the curved length LR and the chord length LR is small, the deformations of the arc LR may be considered approximately equal to the deformations of the chord LR. The accuracy of this approximation is considered in paragraph 5.2. The deformations of the chord LR are known from the solution of the differential equations for the straight beam.

The geometrical relationships between the angles of rotation θ and the bending slopes $\frac{dUy}{dz}$ of the straight and curved elements respectively are expressed by the following equations:

$$\theta_s = \theta_c \cos\alpha - \left(\frac{dUy}{dz}\right)_c \sin\alpha$$

$$\left(\frac{dUy}{dz}\right)_s = \theta_c \sin\alpha + \left(\frac{dUy}{dz}\right)_c \sin\alpha$$

The forces acting at the end R of the arc LR resolved vectorially with respect to the chord LR. The relations between the forces for the curved and straight elements at the end R are referring to Fig. 6 as follows:

$$Ns = Nc \cos\alpha - Qx_c \sin\alpha$$

$$Qx_s = Nc \sin\alpha + Qx_c \cos\alpha$$

$$Ts = Tc \cos\alpha - Mx_c \sin\alpha$$

$$Mx_s = Tc \sin\alpha + Mx_c \cos\alpha$$

The subscripts s and c refer to "straight" and "curved" respectively.

Or in Matrix form

$$
\begin{bmatrix} \theta \\ \dfrac{Uy'}{L} \\ N \\ Qx \\ T \\ Mx \end{bmatrix}_s
=
\begin{bmatrix}
\cos\alpha & -\sin\alpha & 0 & 0 & 0 & 0 \\
\sin\alpha & \cos\alpha & 0 & 0 & 0 & 0 \\
0 & 0 & \cos\alpha & -\sin\alpha & 0 & 0 \\
0 & 0 & \sin\alpha & \cos\alpha & 0 & 0 \\
0 & 0 & 0 & 0 & \cos\alpha & -\sin\alpha \\
0 & 0 & 0 & 0 & \sin\alpha & \cos\alpha
\end{bmatrix}
\begin{bmatrix} \theta \\ \dfrac{Uy'}{L} \\ N \\ Qx \\ T \\ Mx \end{bmatrix}_c
$$

The 6 x 6 Matrix on the right hand side of the above equations may be called "Direction Matrix" D_{cs}.

In a general case the Matrix D will be of order 15 x 15. All the rows, excepting the above six, will have Unity on the column of the diagonal and zeroes on the remaining columns.

The relationsbetween the State Vectors at the end R for the curved and straight element can be written as:

$$\left[S_R\right]_s = \left[D_{cs}\right]\left[S_R\right]_c \qquad\qquad \text{---- (49)}$$

From equation (45) the State Vector at the end L of the straight elements LR is obtained by the relation

$$\left[S_L\right]_s = \left[TR_S\right]_R^L \left[S_R\right]_s \qquad\qquad \text{---- (50)}$$

The State Vector $\left[S_L\right]_c$ for the curved element at the end L can be written as:

$$\left[S_L\right]_c = \left[D_{sc}\right]\left[S_L\right]_s \qquad\qquad \text{---- (51)}$$

Using equations (49) and (50) equation (51) is expressed as follows:

$$\left[S_L\right] = \left[D_{SC}\right]\left[TR_S\right]_R^L\left[D_{CS}\right]\left[S_R\right]_C$$
$$\left[TR_C\right]_R^L\left[S_R\right]_C \qquad \text{---} \quad (52)$$

where the Transport Matrix for the curved element is given by:

$$\left[TR_C\right]_R^L = \left[D_{SC}\right]\left[TR_S\right]_R^L\left[D_{CS}\right] \qquad \text{---} \quad (53)$$

Thus the basic scheme for the solution of the problem of bending and torsion of thin walled beams of open section is the same for both straight and curved beams. In the case of straight beams the Direction Matrices D are Unit Matrices. This procedure has a distinct advantage in the analysis of curved beams with unsymmetrical open section. No general solution for this problem is known. By dividing the length of the curved beam into a sufficiently large number of finite Elements, a sufficient accuracy in practical calculation can be obtained.

3.3 Beams of Variable Section
____

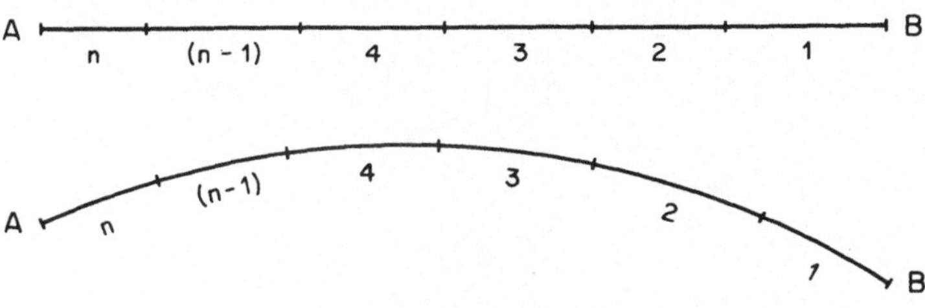

FIG. 7

Fig. 7 shows a beam AB with a straight or curved axis. The
cross section of the beam varies gradually between A and B.
The beam is divided into n elements and each of these elements is assumed to be of constant cross section.

In this case the Transport Matrix form B to A is obtained
as follows:

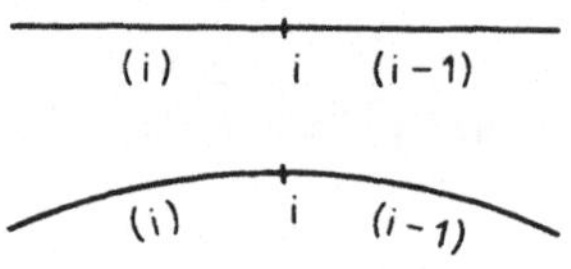

FIG. 8

Fig. 8 shows two neighbouring Finite Elements (i - 1) and
(i) with their junction at section i. The sectional properties of the (i - 1)th and the (i)th element are different.
At i the Section Forces of both elements must be in equilibrium and the deformations must be compatible. In expres-
sing these conditions, for easier computation relative
cross sectional properties will be introduced.

The State Vectors $[S_i]_R$ and $[S_{i-1}]_L$ are premultiplied by
scaling Matrices $[SC]_i$ and $[SC]_{i-1}$ respectively. The equilibrium of forces and compatibility of deformations at i is
expressed by the relation:

$$[SC]_i [S]_i = [SC]_{i-1} [S]_{i-1} \quad\text{---}\quad (54)$$

The scaling Matrices $\begin{bmatrix} SC \end{bmatrix}$ have the following form:

$$[SC] = \begin{bmatrix} 1 \\ & 1 \\ & & 1 \\ & & & 1 & & & & & & & & & O \\ & & & & 1 \\ & & & & & 1 \\ & & & & & & 1 \\ & & & & & & & n_1 \\ & & & & & & & & n_2 \\ & & & & & & & & & n_3 \\ & & & & & & & & & & n_4 \\ & O & & & & & & & & & & n_5 \\ & & & & & & & & & & & & n_6 \\ & & & & & & & & & & & & & n_7 \\ & & & & & & & & & & & & & & 1 \end{bmatrix}$$

The coefficients n_1 to n_7 on the diagonal are dimensionless
parameters as follows:

$$n_1 = \frac{F}{Fm}$$

$$n_2 = n_3 = \frac{Iw}{Iw_m}$$

$$n_4 = n_7 = \frac{Iy}{Iy_m}$$

$$n_5 = n_6 = \frac{Ix}{Ix_m}$$

The subscript m refers to some chosen section. The Transport Matrices between neighbouring sections are obtained
by using the method outlined in paragraph 3.1 and 3.2.
Using equation (54) successively the elements of the Transport Matrix between B and A are calculated.

3.4 The Conditions of Transfer at the Junction of the

Finite Elements

In the following discussion it is assumed that the system made up of the two elements has no jumps as in Fig. 9, but it can have kinks as in Fig. 10.

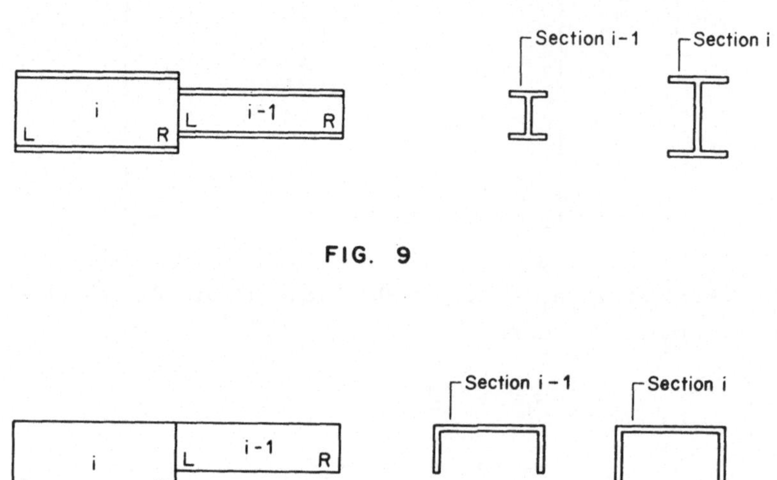

FIG. 9

FIG. 10

Fig. 9 represents an approximation of a tapered I beam. In this case, the normal stresses in the tapered flanges are not parallel to the z axis of the beam and the expression (27) is not valid. A relevant discussion is found in reference [8].

Fig. 10 represents an approximation of a channel section with variable flange width placed sidewise. In this case, when the variation of section along the beam length is gradual there is no great discontinuity in the state of stress on the two sides where the cross section changes.

a) Compatibility of Deformations

Referring to Fig. 10 the conditions of campatibility of deformations on the left and right hand sides of junction i are:

$$1) \quad \theta_L = \theta_R$$

$$2) \quad \theta'_L = \theta'_R$$

$$3) \quad Ux_L = Ux_R$$

$$4) \quad Ux'_L = Ux'_R$$

$$5) \quad Uy_L = Uy_R$$

$$6) \quad Uy'_L = Uy'_R$$

$$7) \quad Uz_L = Uz_R$$

Conditions 3, 5, and 7 express that the displacements of the coordinate origin O which is common to both the left and right hand side sections at i are equal. Equations (4) and (6) state the equalitiy of slopes of the tangents to the deflection curves in the planes of bending xz and yz. Condition (1) expresses that there is no relative angle of twist on the left and right hand sides of junction i. For any other point P which is common to both the elements the displacement parallel to the z axis is given by the relation

$$Uz_p = Uz - \frac{Ux'}{L} \cdot x - \frac{Uy'}{L} \cdot y - \frac{\theta'}{L} \cdot w \quad —— (55)$$

For equal x, y, and w by virtue of conditions 2, 4, and 6 the displacement Uz_p is equal for both Finite Elements.

The compatibility of deformations is thus completely
defined by relations 1 to 7 given at the beginning of the
paragraph.

b) Equilibrium of Forces

This is represented in Matrix equation (54) by the conditions:

$$8) \quad N_L = N_R$$

$$9) \quad T_L = T_R$$

$$10) \quad Mw_L = Mw_R$$

$$11) \quad My_L = My_R$$

$$12) \quad Mx_L = Mx_R$$

$$13) \quad Qx_L = Qx_R$$

$$14) \quad Qy_L = Qy_R$$

If concentrated forces are acting at the junction i, they
can be combined to build a Load Vector $[P]_i$ and the effect
of this Load Vector on the forces and deformations at the
end L is obtained by premultiplying it by the Transport
Matrix between i and L. If the loads act along the positive
directions of the axes, their signs are taken as negative.

c) Continuous Beams - Conditions of Transfer at the intermediate Supports

An intermediate support of a continuous beam is shown in
Fig. 11. The beam is free to rotate in the yz and xz planes
at this section, but it is prevented from rotating in the
plane xy by supports as shown in the figure.

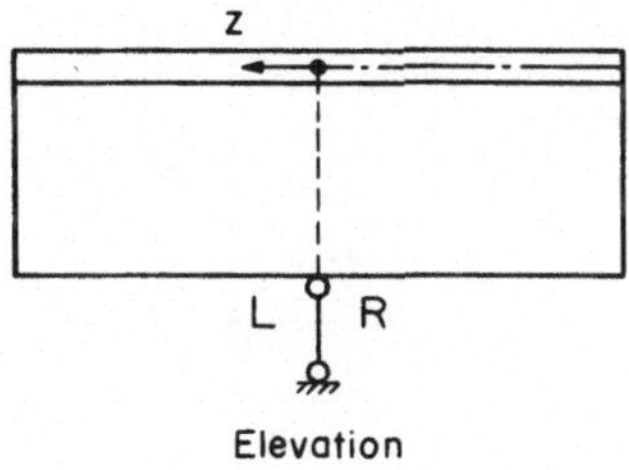

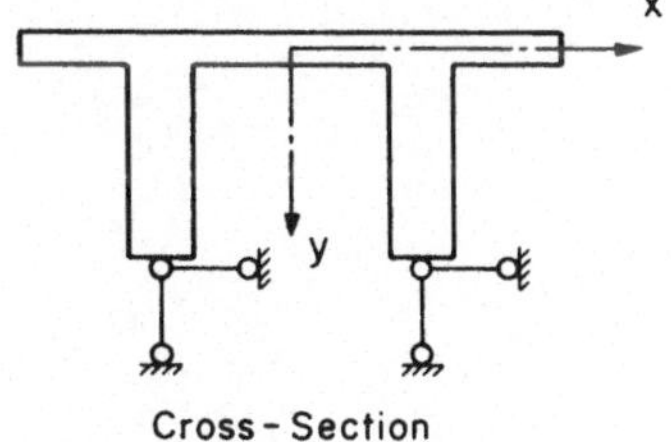

Elevation   Cross-Section

FIG. 11

The conditions of transfer are in this case as follows:

Geometrical Conditions

$$
\begin{aligned}
1)\quad & \theta_L && = && 0 \\
2)\quad & \theta_R && = && 0 \\
3)\quad & Ux_L && = && 0 \\
4)\quad & Ux_R && = && 0 \\
5)\quad & Uy_L && = && 0 \\
6)\quad & Uy_R && = && 0 \\
7)\quad & Uz_L && = && Uz_R \\
8)\quad & Ux'_L && = && Ux'_R \\
9)\quad & Uy'_L && = && Uy'_R \\
10)\quad & \theta'_L && = && \theta'_R
\end{aligned}
$$

<u>Statical Conditions</u>

$$11) \quad N_L \quad = \quad N_R$$

$$12) \quad Mx_L \quad = \quad Mx_R$$

$$13) \quad My_L \quad = \quad My_R$$

$$14) \quad Mw_L \quad = \quad Mw_R$$

For matrix operation, it is more convenient to express the previous conditions in the following form:

$$
14 \text{ Equations} : \quad
\begin{bmatrix}
\theta \\
\theta' \\
Ux \\
Ux' \\
Uy \\
Uy' \\
Uz \\
N \\
T \\
Mw \\
My \\
Mx \\
Qx \\
Qy
\end{bmatrix}_L
=
\begin{bmatrix}
\theta \\
\theta' \\
Ux \\
Ux' \\
Uy \\
Uy' \\
Uz \\
N \\
T \\
Mw \\
My \\
Mx \\
Qx \\
Qy
\end{bmatrix}_R
-
\begin{bmatrix}
O \\
O \\
O \\
O \\
O \\
O \\
O \\
O \\
T \\
O \\
O \\
O \\
Rx \\
Ry
\end{bmatrix}
$$

In the above vector equation three new unknowns are artifi-
cially introduced, namely the support reactions T,. Rx, and
Ry. These are then obtained by using the conditions

$$\theta = 0$$

$$3 \text{ Equations}: \quad Ux = 0$$

$$Uy = 0$$

Thus three unknowns and three conditions are introduced pur-
posely to facilitate the Matrix operations. This procedure
is used in the computer program discussed in chapter 4
to analyse continuous beams.

The number of conditions of transfer at each junction of
the Finite Elements remains fourteen in every case.

c) Skew Supports

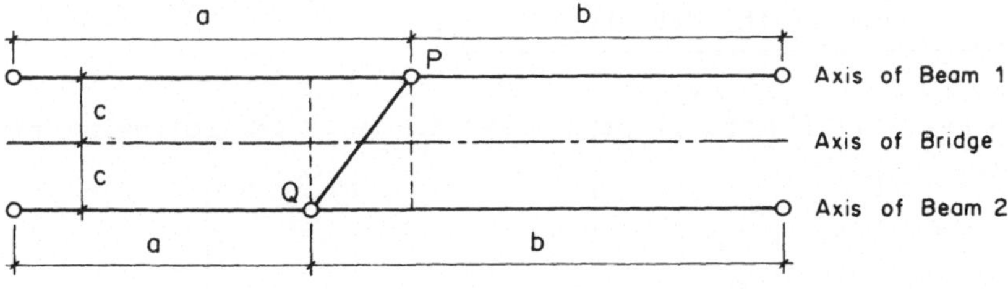

FIG. 12

Fig. 12 shows a two beam bridge in plan. The intermeditate
support PQ is skew to the axis of the bridge. P and Q may
be treated in this case as two point supports. The conditions at cross sections P and Q may be written as follows:

$$\text{Section P}: \quad Ur + c \cdot \theta = 0 , \quad Ux - h\theta = 0$$

$$\text{Section Q}: \quad Ur - c \cdot \theta = 0 , \quad Ux - h\theta = 0$$

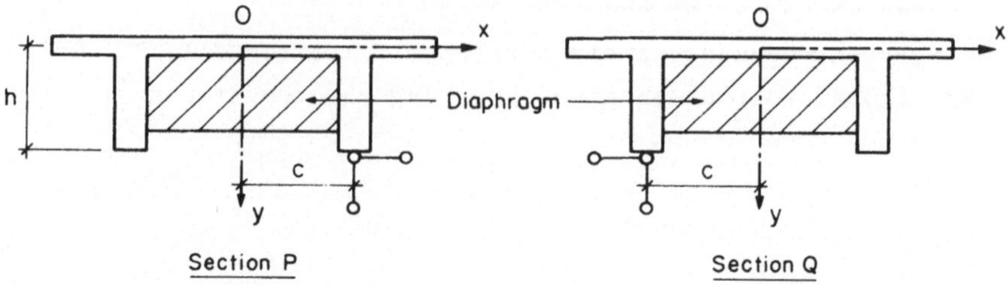

FIG. 13

As discussed in part c) of this paragraph the above conditions enable to determine the unknown vertical reactions at
P and Q which are in this case eccentric to the axis O_y.

In practise the condition that the cross sectional shape in
sections P and Q remains unchanged requires diaphragms as
shown in Fig. 13.

3.5 Nature of the Transport Matrix

On partitioning the Transport Matrix shows the following form:

$$
\begin{bmatrix}
\begin{array}{c|c}
\begin{matrix}[TR]_D \\ 7 \times 7\end{matrix} & \begin{matrix}[F] \\ 7 \times 7\end{matrix} \\
\hline
\begin{matrix}[S] \\ 7 \times 7\end{matrix} & \begin{matrix}[TR]_F \\ 7 \times 7\end{matrix} \\
\hline
\multicolumn{2}{c}{\begin{matrix}[O] \\ (1 \times 14)\end{matrix}}
\end{array}
\end{bmatrix}
\quad \begin{bmatrix}L_o\end{bmatrix} \ (14 \times 1)
$$

The submatrices $[TR]_D$ and $[TR]_F$ represent the "Deformation Transport Matrix" and the "Force Transport Matrix" respectively. Their elements depend on the geometry of the system that is analysed. $[F]$ is the Flexibility Matrix and $[S]$ the Stiffness Matrix. The Vector $[Lo]$ contains the loading terms.

4. COMPUTER PROGRAM FOR THE ANALYSIS OF THIN WALLED

BEAMS OF VARIABLE SECTION

Based on the theory outlined in chapter 2 and 3, a computer
program was written in the "ALGOL" language and the numerical examples described in chapter 5 were worked out by using
this program on the CDC 1604 A Computer of the Computation
Centre, Swiss Federal Institute of Technology, Zurich. This
program is abailable in the library of the Computation
Centre. A complete description will not be given here. A
scheme of the program is given on the following page.

SCHEME OF PROGRAM

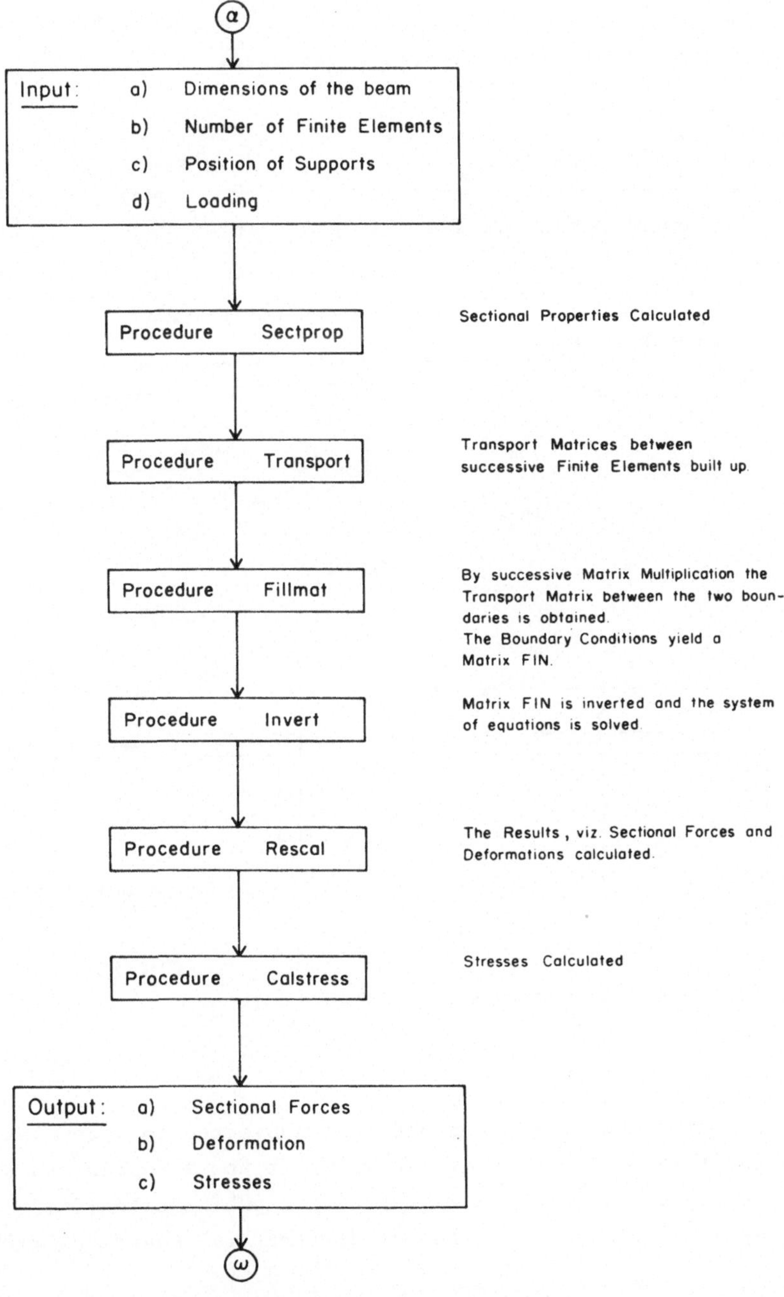

5. NUMERICAL EXAMPLES

The theory outlined in chapters 2, 3, and 4 will now be
illustrated by some numerical examples.

5.1 Fixed Ended Beam with a Concentrated Torsional

Moment Acting at the Mid-Span Section

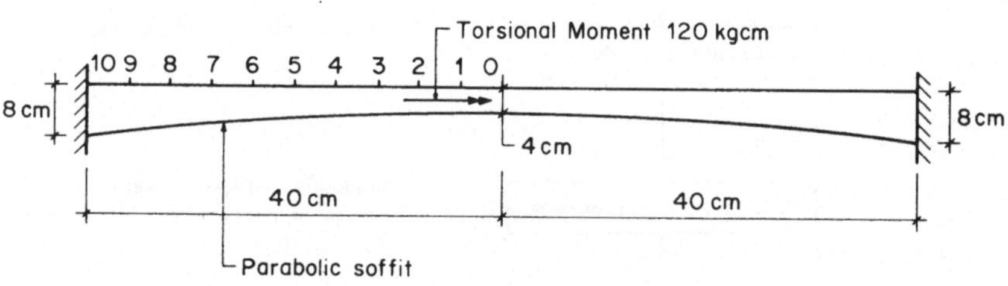

FIG. 14

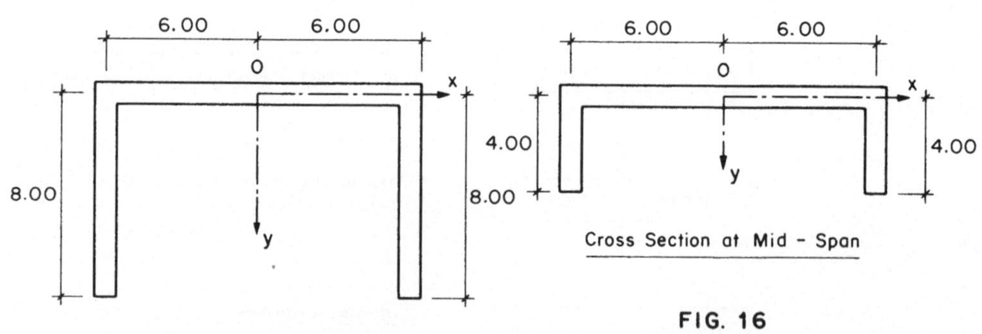

FIG. 15

Fig. 14 shows a fixed ended beam subjected to a concentrated torsional moment of 120 kg cm at the mid-span cross
section. This example is taken from publication $[2]$.
Cywinski obtains a system of differential equations with

variable coefficients for the torsion of a beam of variable section, and he solves it by finite differences. The same problem is now solved by finite elements and the two solutions are compared.

The left half of the beam is divided into ten Finite Elements, each of constant cross section. The properties of these cross sections are given in tables 1 and 2. The symmetry of the beam and of the loading is taken onto account. Accordingly, at the right of Element 1 we have the following boundary conditions.

$$
\begin{aligned}
1)\quad & U_x{}' = 0 \\
2)\quad & U_y{}' = 0 \\
3)\quad & \theta{}' = 0 \\
4)\quad & T = -\frac{120}{2} = -60 \\
5)\quad & N = 0 \\
6)\quad & Q_x = 0 \\
7)\quad & Q_y = 0
\end{aligned}
$$

At the left of element 10 the conditions are those for a fixed end. The results obtained are tabulated in tables 3 and 4.

Table 1 - Sectional Properties of the Finite Elements

Element	F_{cm^2}	Sx_{cm^3}	Gy_{cm}	Ky_{cm}	Ix_{cm^4}	Iy_{cm^4}
1	12.01	9.65	0.80	- 1.34	25.8	259.5
2	12.11	10.03	0.83	- 1.37	27.4	263.0
3	12.30	10.83	0.88	- 1.45	30.7	269.9
4	12.59	12.09	0.96	- 1.55	36.2	280.3
5	12.97	13.88	1.07	- 1.70	44.5	294.2
6	13.45	16.29	1.21	- 1.88	56.6	311.4
7	14.03	19.43	1.38	- 2.11	73.7	332.2
8	14.70	23.43	1.59	- 2.37	97.6	356.4
9	15.46	28.48	1.84	- 2.67	130.8	384.0
10	16.33	34.74	2.12	- 3.01	176.2	415.1

Table 2 - Sectional Properties (Contd.)

Element	Iwx_{cm^5}	Iw_{cm^6}	$I\bar{x}_{cm^4}$	$I\bar{y}_{cm^4}$	$I\bar{w}_{cm^6}$	K_{cm^4}
1	347.3	928.0	18.0	259.5	463.0	1.44
2	361.3	985.0	19.0	263.0	488.0	1.45
3	390.2	1105.0	21.2	269.9	541.0	1.48
4	435.4	1303.0	24.6	280.3	626.0	1.51
5	499.7	1602.0	29.6	294.2	753.0	1.56
6	586.3	2036.0	36.8	311.4	932.0	1.61
7	699.3	2652.0	46.8	332.2	1179.0	1.68
8	843.8	3514.0	60.3	356.4	1517.0	1.76
9	1025.4	4708.0	78.4	384.0	1970.0	1.86
10	1250.9	6344.0	102.3	415.1	2574.0	1.96

Table 3 - Deformations of the Beam under a concentrated
Torsional Moment of 120 kg/cm

Section	Finite Difference Solution		Finite Elements Solution	
	Θ $(10^{-4}$ Rad.$)$	Ux $(10^{-4}$ cm$)$	Θ $(10^{-4}$ Rad.$)$	Ux $(10^{-4}$ cm$)$
0	90.74	- 172.28	89.18	- 169.19
1	86.29	- 165.49	85.20	- 163.51
2	75.79	- 148.84	75.15	- 147.41
3	61.99	- 125.87	61.71	- 124.93
4	47.22	- 99.83	47.21	- 99.29
5	33.23	- 73.01	33.37	- 73.26
6	21.17	- 49.19	21.40	- 49.13
7	11.69	- 28.60	11.92	- 28.65
8	5.04	- 13.02	5.21	- 13.11
9	1.20	- 3.31	1.29	- 3.38
10	0	0	0	0

Table 4 - Sectional Forces (Example 1)

Section	Finite Difference Solution					Finite Elements Solution					
	T_s	T_w	T	$M_y = M_{\bar{y}}$	$M_{\bar{w}}$	T_s	T_w	T	$M_y = M_{\bar{y}}$	M_w	$M_{\bar{w}}$
	kgcm	kgcm	kgcm	kgcm	$kgcm^2$	kgcm	kgcm	kgcm	kgcm	$kgcm^2$	$kgcm^2$
0	0	59.78	59.78	- 83.22	769.6	0	- 60.00	- 60.00	81.94	665.1	775.1
1	2.95	57.30	60.25	- 83.22	536.6	- 2.96	- 57.04	- 60.00	81.94	431.3	541.4
2	4.85	55.38	60.23	- 83.22	316.7	- 4.84	- 55.16	- 60.00	81.94	207.3	320.2
3	5.81	54.37	60.18	- 83.22	105.5	- 5.77	- 54.23	- 60.00	81.94	- 110.2	107.7
4	6.01	54.13	60.14	- 83.22	- 100.3	- 5.93	- 54.07	- 60.00	81.94	- 227.1	- 99.5
5	5.63	54.47	60.10	- 83.22	- 303.3	- 5.53	- 54.47	- 60.00	81.94	- 443.6	- 304.0
6	4.84	35.23	60.07	- 83.22	- 505.5	- 4.73	- 55.23	- 60.00	81.94	- 662.4	- 507.8
7	3.79	56.25	60.04	- 83.22	- 708.0	- 3.69	- 56.31	- 60.00	81.94	- 885.1	- 712.2
8	2.59	57.43	60.02	- 83.22	- 911.7	- 2.51	- 57.49	- 60.00	81.94	- 1112.2	- 917.8
9	1.31	58.70	60.01	- 83.22	- 1117.1	- 1.27	- 58.73	- 60.00	81.94	- 1344.2	- 1125.0
10	0	60.00	60.00	- 83.22	- 1324.2	0	- 60.00	- 60.00	81.94	- 1581.4	- 1334.2

In the second part of the problem we consider a beam of
unsymmetrical section. The web on the right hand side of
the axis Oy is now longer than the one on the left hand
side as shown in Fig. 17. The other dimensions remain the
same as in the first part of the problem. See tables 5 to 8.

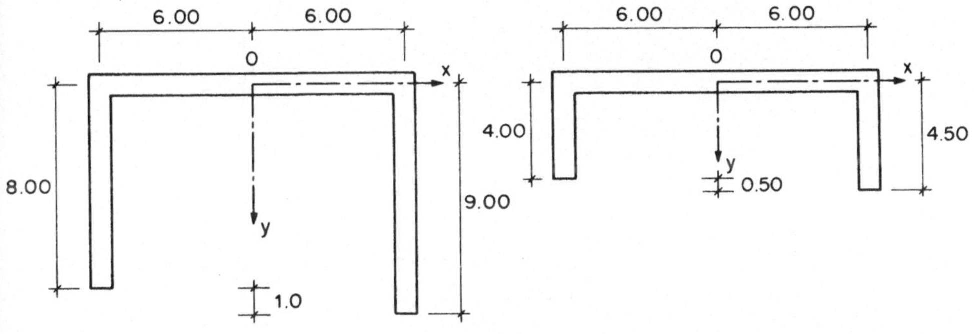

FIG. 17

Table 5 - Sectional Properties (Fig. 17)

Element	F_{cm^2}	Sx_{cm^3}	Sy_{cm^3}	Sw_{cm^4}	Gx_{cm}	Gy_{cm}	Kx_{cm}	Ky_{cm}
1	12.31	10.93	1.80	7.69	0.147	0.887	0.696	- 1.435
2	12.41	11.37	1.84	8.00	0.148	0.916	0.694	- 1.473
3	12.62	12.28	1.91	8.64	0.152	0.972	0.689	- 1.549
4	12.92	13.70	2.02	9.64	0.156	1.060	0.683	- 1.664
5	13.33	15.73	2.16	11.06	0.162	1.179	0.676	- 1.819
6	13.84	18.45	2.34	12.98	0.169	1.333	0.667	- 2.014
7	14.45	22.01	2.56	15.48	0.177	1.522	0.659	- 2.250
8	15.17	26.55	2.81	18.68	0.185	1.750	0.650	- 2.528
9	15.98	32.27	3.10	22.70	0.194	2.019	0.641	- 2.849
10	16.90	39.36	3.42	27.69	0.203	2.329	0.632	- 3.212

Table 6 - <u>Sectional Properties (Unsymmetrical Section)</u>

(In Continuation of Table 5)

Element	Ix_{cm^4}	Iy_{cm^4}	Ixy_{cm^4}	Iwx_{cm^5}	Iwy_{cm^5}	Iw_{cm^6}	$I\bar{x}_{cm^4}$	$I\bar{y}_{cm^4}$	$I\bar{w}_{cm^6}$	K_{cm^4}
1	31.25	270.4	7.69	393.5	32.78	1124.8	21.55	270.2	537.3	1.477
2	33.17	274.1	8.00	409.3	34.78	1193.5	22.75	273.8	566.4	1.490
3	37.21	281.4	8.64	442.0	39.03	1339.1	25.27	281.2	627.5	1.514
4	43.88	292.5	9.64	493.2	46.02	1579.1	29.35	292.1	726.7	1.551
5	53.94	307.1	11.06	566.1	56.58	1941.3	35.39	306.8	873.4	1.600
6	68.55	325.5	12.98	664.2	71.90	2467.0	43.96	325.1	1081.4	1.661
7	89.29	347.5	15.48	792.2	93.66	3213.6	55.79	347.1	1369.4	1.734
8	118.34	373.2	18.68	955.8	124.10	4258.9	71.86	372.7	1761.7	1.820
9	158.54	402.6	22.70	1161.6	166.30	5705.8	93.41	402.0	2289.9	1.918
10	213.62	435.7	27.69	1417.0	224.10	7688.0	121.95	435.0	2994.2	2.028

Table 7 - Deformations under a concentrated Torsional Moment of 120 kg/cm at the Mid-Span Section

Section	Θ (10^{-4} Rad.)	Ux (10^{-4} cm)	Uy (10^{-4} cm)	Uz (10^{-4} cm)
0	77.60	- 157.7	- 62.5	0
1	74.14	- 152.0	- 59.7	0.61
2	65.38	- 137.0	- 52.7	1.04
3	53.69	- 116.0	- 43.2	1.31
4	41.05	- 92.1	- 33.1	1.42
5	29.01	- 68.0	- 23.4	1.40
6	18.59	- 45.5	- 15.0	1.27
7	10.35	- 26.5	- 8.3	1.05
8	4.35	- 12.1	- 3.7	0.75
9	1.12	- 3.1	- 0.9	0.39
10	0	0	0	0

Table 8 - Sectional Forces

Fixed Ended Beam of Variable Unsymmetrical Section under a concentrated Torsional Moment of 120 kg cm at Mid-Span

Section	T_s kgcm	T_w kgcm	T kgcm	M_y kgcm	M_x kgcm	N kg	M_w kgcm^2	$\bar{M_w}$ kgcm^2
0	0	- 60.00	- 60.00	78.71	- 1.47	0.523	665.8	780.2
1	- 2.64	- 57.36	- 60.00	78.71	- 1.47	0.523	431.4	545.8
2	- 4.32	- 55.68	- 60.00	78.71	- 1.47	0.523	205.7	323.0
3	- 5.15	- 54.85	- 60.00	78.71	- 1.47	0.523	- 15.0	108.3
4	- 5.30	- 54.70	- 60.00	78.71	- 1.47	0.523	- 233.6	- 101.2
5	- 4.95	- 55.05	- 60.00	78.71	- 1.47	0.523	- 452.6	- 308.0
6	- 4.23	- 55.77	- 60.00	78.71	- 1.47	0.523	- 673.7	- 513.8
7	- 3.30	- 56.70	- 60.00	78.71	- 1.47	0.523	- 898.1	- 719.6
8	- 2.25	- 57.75	- 60.00	78.71	- 1.47	0.523	- 1126.5	- 926.1
9	- 1.14	- 58.86	- 60.00	78.71	- 1.47	0.523	- 1359.4	- 1133.8
10	0	- 60.00	- 60.00	78.71	- 1.47	0.523	- 1596.9	- 1342.7

Discussion of the Results

The values in tables 3 and 4 show very good agreement between the values of the deformations and sectional forces as given by the finite difference solution and the finite elements solution respectively. Again the values in tables 3 and 7 and 4 and 8 respectively may be compared to see how the theory outlined in the present work gives a good solution for torsion of beams of non-symmetrical variable section.

Comparison of the Finite Differences and Finite Elements Solution

For the general case of the torsion of a beam of variable section Cywinski [2] obtains a set of four differential equations with variable coefficients. Using the notation of the present thesis, and for the case $q_x = q_y = q_z = 0$, these equations are written down as:

$$L\,(EF\,U_z')' - (ES_yU_x'')' - (ES_xU_y'')' - (ES_w\theta'')' = 0$$

$$-L\,(ES_yU_z')'' + (EI_yU_x'')'' + (EI_{xy}U_y'')'' + (EI_{wx}\theta'')'' = 0$$

$$-L\,(ES_xU_z')'' + (EI_{xy}U_x'')'' + (EI_xU_y'')'' + (EI_{wy}\theta'')'' = 0$$

$$-L\,(ES_wU_z')'' + (EI_{wx}U_x'')'' + (EI_{wy}U_y'')'' + (EI_w\theta'')'' - L^2(GK\theta')' = mL^4$$

The above equations exactly represent the equilibrium of the element of variable section under the external torsional moment m. The approximate solutions of these equations are obtained numerically. In the case discussed in paragraph 3.3, the approximation lies in the physical idealisation. The solutions of the differential equations are exact.

5.2 Three Span Continuous Curved Beam

Fig. 18 shows a three span continuous beam 160 meters long.
It is loaded at the mid-span section with a load of 100 tons
acting in the radial direction. The axis of the beam is an
arc of a circle of 400 meters radius. This numerical data
is taken from reference [3]. This choice enables us to com-
pare the results obtained for a particular problem of open
profile curved girders from two different approaches,
namely:

1) Differential Equation of Torsion of a curved beam

2) Differential Equation of Torsion of a straight beam

In paragraph 4.3 it has already been explained, as to how
the Transport Matrices for curved elements are obtained
within the scope of the present theory.

The results are given in Tables 9 to 12.

The beam is divided into 16 elements as shown in Fig. 18,
and the symmetry of the beam and of the loading is taken
into account.

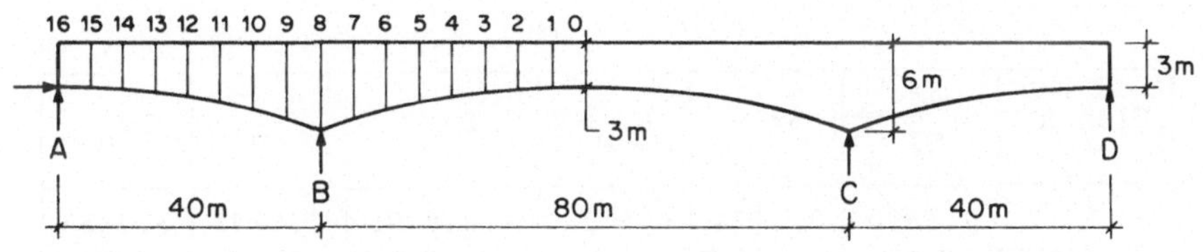

A) ELEVATION

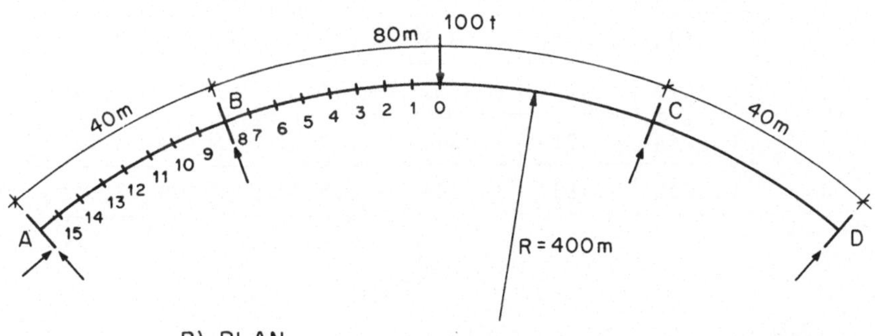

B) PLAN

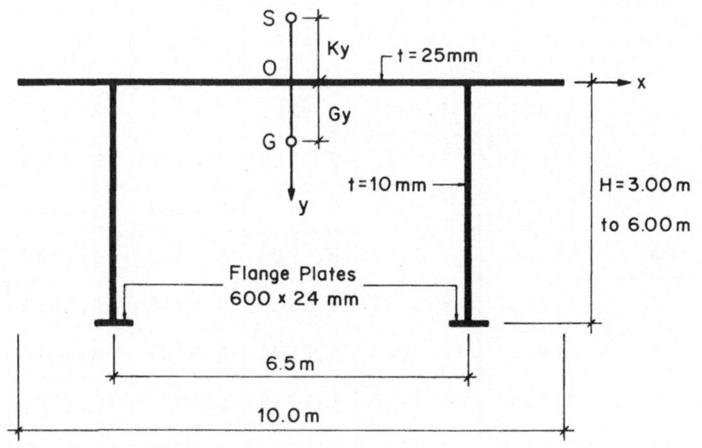

C) CROSS SECTION

FIG. 18

Table 9 - Sectional Properties (Example 2)

Element	F_{m^2}	Sx_{m^3}	Gy_m	Ky_m	Ix_{m^4}	Iy_{m^4}
1	0.3390	0.1774	0.5234	- 0.6198	0.4433	3.0238
2	0.3409	0.1859	0.5452	- 0.6452	0.4774	3.0436
3	0.3447	0.2033	0.5838	- 0.6954	0.5503	3.0832
4	0.3503	0.2307	0.6586	- 0.7754	0.6723	3.1426
5	0.3578	0.2697	0.7538	- 0.8842	0.8598	3.2218
6	0.3672	0.3224	0.8782	- 1.0255	1.1370	3.3209
7	0.3784	0.3915	1.0346	- 1.2022	1.5379	3.4397
8	0.3915	0.4801	1.2261	- 1.4171	2.1089	3.5783

Table 10 - Sectional Properties (Contd.)

Element	Iwx_{m^5}	Iw_{m^6}	$I\bar{x}_{m^4}$	$I\bar{y}_{m^4}$	$I\bar{w}_{m^6}$	K_{m^4}
1	1.8742	4.6821	0.3505	3.0238	3.5204	0.0001
2	1.9633	5.0417	0.3760	3.0436	3.7753	0.0001
3	2.1471	5.1821	0.4304	3.0832	4.3169	0.0001
4	2.4366	7.1002	0.5204	3.1426	5.2109	0.0001
5	2.8487	9.0779	0.6565	3.2218	6.5611	0.0001
6	3.4056	12.0072	0.8538	3.3209	8.5148	0.0001
7	4.1351	16.2416	1.1329	3.497	11.2705	0.0001
8	5.0707	22.2713	2.5203	3.5783	15.0859	0.0001

Table 11 - Deformations (Example 2)

Section	Θ 10^{-4} Rad.	Θ * 10^{-4} Rad.	Ux 10^{-4} cm	Uy 10^{-4} cm
(Mid-Span) 0	- 29.28	- 28.19	9474	61.6
1	- 28.24	- 27.19	9138	59.5
2	- 25.50	- 24.54	8363	53.8
3	- 21.60	- 20.78	7171	45.6
4	- 17.04	- 16.39	5729	36.0
5	- 12.26	- 11.80	4166	25.9
6	- 7.65	- 7.36	2608	16.1
7	- 3.49	- 3.36	1179	7.4
8 Intermediate Support	0	0	0	0
9	2.64	2.54	- 838	- 5.6
10	4.42	4.26	- 1349	- 9.4
11	5.36	5.15	- 1574	- 11.4
12	5.48	5.27	- 1557	- 11.6
13	4.86	4.67	- 1343	- 10.3
14	3.62	3.48	- 979	- 7.7
15	1.93	1.85	- 515	- 4.1
16 Left End	0	0	0	0

* Reference   3

Table 12 - Sectional Forces (Example 2)

Section	N t	Q_x t	M_y mt	M_w tm^2	$\overline{M_w}$ tm^2	M_w * tm^2
0	- 3.20	- 50.00	- 1279	129.08	- 663.9	- 643.1
1	- 2.57	- 50.04	- 1029	129.07	- 508.9 - 534.8	- 489.2 - 515.1
2	- 1.95	- 50.06	- 779	129.04	- 373.4 - 413.4	- 356.9 - 396.6
3	- 1.32	- 50.08	- 529	129.00	- 239.0 - 280.8	- 227.7 - 269.0
4	- 0.70	- 50.10	- 278	128.94	- 86.7 - 117.0	- 82.4 - 112.2
5	- 0.07	- 50.10	- 28	128.87	104.5 100.5	99.6 96.3
6	0.56	- 50.10	223	128.77	357.3 396.8	341.4 381.5
7	1.18	- 50.09	474	128.66	697.7 799.5	668.7 771.3
8	1.81	- 50.07 18.03	724	128.53	1154.2	1110.0
9	1.58	18.06	634	112.50	1010.3 874.1	976.3 839.4
10	1.36	18.09	543	96.46	749.5 653.6	723.2 626.6
11	1.13	18.10	453	80.40	544.8 480.8	527.8 460.4
12	0.91	18.11	362	64.34	384.7 345.3	370.0 330.3
13	0.68	18.12	272	48.26	259.0 237.5	248.2 227.2
14	0.45	18.12	181	32.18	158.4 149.1	151.9 142.6
15	0.23	18.12	91	16.09	74.6 72.2	71.4 69.1
16	0	18.12	0	0	0	0

* Reference 3

Discussion of the Results

The comparison of the second and third columns in table 11 shows us the range of occuracy of the approach outlined in the present work. The values of the angle of twist are 3,5 % higher than the values given by Becker. The warping moment is a function of the second derivative of θ and consequently the values of the warping moment with respect to an axis perpendicular to the plane of cross section and passing through the shear centre, which the present theory yields are also 3,5 % higher than those given by Becker.

This difference arises from the fact that in a curved beam, the torsional moment and the bending moment, which are the tangential and normal components of the moment vector at a cross section, are continuously resolved. In the present theory the circular arc is replaced by a chain of chords and the vector resolution is carried out only at the junctions of these chords. The closer the chords approximate the arcs, the greater would be the accuracy. Yet an accuracy of 96,5 % is sufficient for problems in practise.

The advantage in starting from the fourth order differential equation of a straight beam lies in the fact that for this equation, closed solutions are obtained straight forward, as compared to the series solutions for the sixth order differential equation for a curved beam. As far as is known to the present author, the differential equation of torsion of curved open profile beams has so far been derived only for beams with one axis of symmetry in the cross section. In the present work, the system of differential equations is considered with respect to an arbitrary set of x, y, z axes and is valid for non-symmetrical sections also.

5.3 Numerical Example III

In this example we consider the statical analysis of a two
span continuous prestressed concrete bridge. The spans are
each 60 meters. In cross section the bridge is three beam
and deck-slab type. The central girder has in plan an axis
with a radius of curvature of 100 meters. Fig. 20 shows
schematically the bridge in elevation and Figures 21 and
22 represent the cross sections of the bridge at the end
and central supports respectively. The deck slab is 20 cm
thick and the webs of the beam are each 50 cm thick. The
depth of the beams varies from 2,5 meters at the end supports to 4 meters at the central support.

The statical analysis of the bridge is carried out for the
following cases of loading:

   i) Prestressing

  ii) Prestressing and dead load

 iii) Prestressing, dead load and live load (360 kg/m^2)
acting on both spans,

  iv) Prestressing, dead load, and live load (360 kg/m^2)
acting on one span only

   v) Prestressing, dead load, and live load acting on
half the width of the bridge on both spans

  vi) Prestressing, dead load, and live load acting on
half the width of the bridge on one span only.

For the purpose of the illustrative example, the impact
factor is not taken into account.

Prestressing: the prestressing cables are laid symmetrical
in both spans and in each of the three beams they have
the same eccentricity with respect to the axis Ox at each
cross section. See Fig. 19.

The bearings at the ends and at the intermediate support
are assumed to lie on the radius at the respective sections.

The values of the prestressing force in each of the three
cables are as follows:
Outer-Girder: 4000 tons
Central-Girder: 3000 tons
Inner-Girder: 2000 tons

The properties of the different cross sections are tabulated in tables 13 and 14. Since both the spans of the beam
are symmetrical to each other the values are given only for
the properties of sections 0 to 8. For each element the
depth of the bridge is assumed to be constant and equal to
the actual depth at the centre of each segment.

Tables 15 to 20 give the values of the Deformations and the
Sectional Forces in sections 0 to 8.

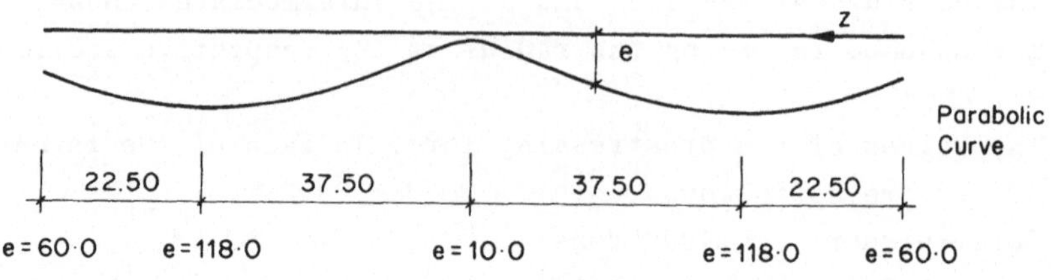

FIG. 19   Profile of the Prestressing Cable

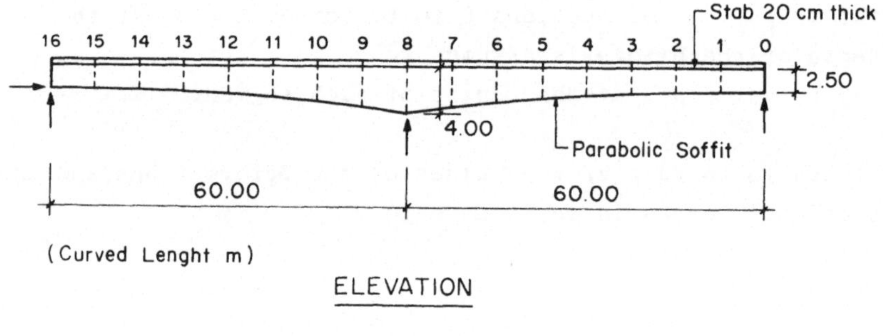

FIG. 20

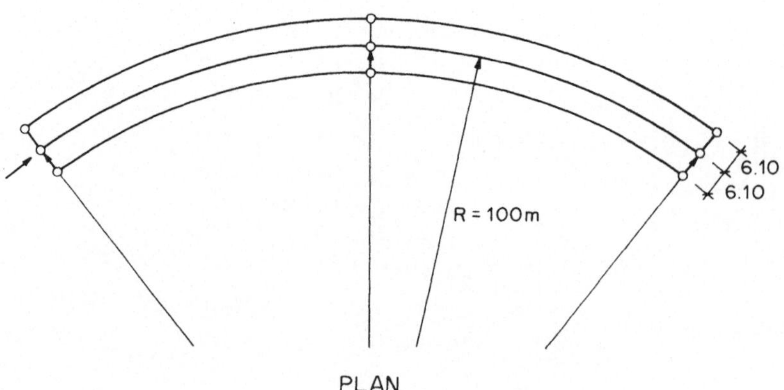

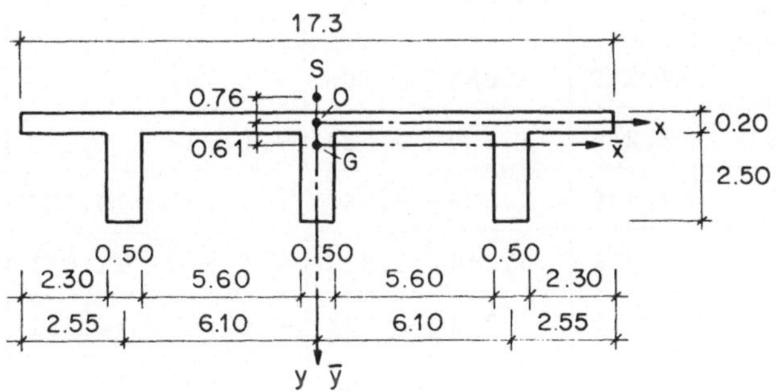

Cross Section – End Support

FIG. 21

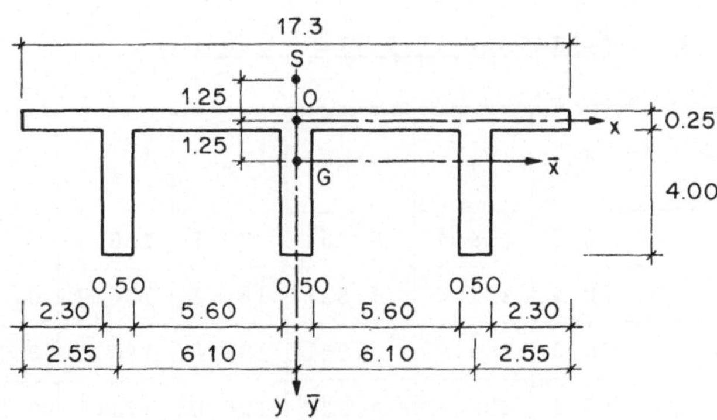

Cross Section – Central Support

FIG. 22

(All dimensions in Meters)

Table 13 - Sectional Properties (Example 3)

Element	F_{m^2}	Sx_{m^3}	Gy_m	Ky_m	Ix_{m^4}	Iy_{m^4}
1	7.219	5.085	0.705	- 0.704	8.846	173.5
2	7.289	5.270	0.723	- 0.722	9.332	181.3
3	7.430	5.650	0.760	- 0.760	10.357	184.8
4	7.641	6.244	0.817	- 0.816	12.031	190.0
5	7.922	7.082	0.894	- 0.893	14.531	197.0
6	8.274	8.205	0.991	- 0.990	18.114	205.7
7	8.695	9.660	1.111	- 1.109	23.136	216.2
8	9.188	11.508	1.253	- 1.251	30.077	228.4

Table 14 - Sectional Properties (Contd.)

Element	Iwx_{m^5}	Iw_{m^6}	$I\bar{x}_{m^4}$	$I\bar{y}_{m^4}$	$I\bar{w}_{m^6}$	K_{m^4}
1	126.3	219.4	5.264	179.5	130.5	0.3593
2	130.9	231.4	5.522	181.3	136.9	0.3652
3	140.3	256.9	6.061	184.8	150.3	0.3769
4	155.1	298.4	6.928	190.0	171.8	0.3945
5	175.9	360.4	8.199	197.0	203.3	0.4179
6	203.7	449.2	9.977	205.7	247.5	0.4472
7	239.8	573.8	12.405	216.2	307.7	0.4824
8	285.7	745.9	15.663	228.4	388.6	0.5234

Table 15 - <u>Angle of Twist Θ (Example 3)</u>

Section	Loading Case I	Loading Case II	Loading Case III	Loading Case IV	Loading Case V	Loading Case VI
0	0	0	0	0	0	0
1	0.31	- 4.65	- 6.17	- 9.3	- 0.82	0
2	0.70	- 7.65	- 10.2	- 15.9	- 1.14	0.45
3	0.99	- 8.33	- 11.2	- 18.6	- 0.81	1.47
4	1.07	- 6.95	- 9.4	- 17.3	- 0.03	2.75
5	0.93	- 4.43	- 6.1	- 13.4	0.70	3.67
6	0.63	- 1.94	- 2.7	- 8.3	0.91	3.59
7	0.27	- 0.36	- 0.6	- 3.6	0.51	2.24
8	0	0	0	0	0	0

Table 16 - <u>Vertical Deflection Uy_0 (Example 3)</u>
(Unit - Millimeters)

Section	Loading Case I	Loading Case II	Loading Case III	Loading Case IV	Loading Case V	Loading Case VI
0	0	0	0	0	0	0
1	- 16.91	2.97	9.10	13.1	5.97	7.95
2	- 29.71	4.12	14.51	22.1	9.21	12.90
3	- 34.92	4.01	15.91	26.0	9.84	14.75
4	- 32.30	3.11	13.89	25.1	8.39	13.86
5	- 23.97	1.87	9.68	20.4	5.70	10.96
6	- 13.36	0.67	4.89	13.6	2.74	7.00
7	- 4.23	- 0.08	1.16	6.2	0.52	3.02
8	0	0	0	0	0	0

Table 17 - Bending Moment $M\bar{x}$ (Example 3)
(Unit - Meter-Tons)

Section	Loading Case I	Loading Case II	Loading Case III	Loading Case IV	Loading Case V	Loading Case VI
0	919	919	919	919	919	919
1	- 1690	688	1433	1643	1053	1156
2	- 3223	508	1668	2088	1077	1282
3	- 3692	353	1596	2224	962	1270
4	- 3105	189	1183	2019	676	1086
5	- 1470	- 25	387	1431	177	688
6	1206	- 340	- 839	410	- 583	28
7	4916	- 811	- 2552	- 1099	- 1663	- 952
8	9650	- 1507	- 4816	- 3162	- 3127	- 2317

Table 18 - Torsional Moments Ts, Tw for different Cases
of Loading (Example 3)

Section	Loading Case I		Loading Case II		Loading Case III	
	Ts	Tw	Ts	Tw	Ts	Tw
0	1.5	6.5	- 38.7	- 199.2	- 51.1	- 262.9
1	2.9	8.8	- 30.7	- 159.1	- 41.0	- 210.7
2	2.8	9.9	- 14.3	- 60.0	19.4	- 81.1
3	1.5	8.0	3.7	64.7	4.5	82.8
4	- 0.3	1.9	16.6	181.8	21.7	237.4
5	- 1.8	- 8.8	20.6	254.4	27.4	334.6
6	- 2.7	- 23.9	16.2	240.9	22.0	320.7
7	- 2.6	- 42.2	7.1	95.5	10.1	136.2
8	- 1.4	- 61.7	- 1.4	- 230.8	- 1.4	- 281.8

Table 18 (Contd.) - Torsional Moments Ts, Tw for different Cases of Loading (Example 3)

Section	Loading Case IV		Loading Case V		Loading Case VI	
	Ts	Tw	Ts	Tw	Ts	Tw
0	- 75.6	- 328.8	- 7.6	- 7.4	- 1.3	- 0.9
1	- 63.5	- 274.7	- 4.7	- 50.4	1.5	- 41.9
2	- 36.3	- 139.0	0.1	- 17.5	5.8	- 2.7
3	- 4.2	36.4	4.8	47.3	9.5	72.7
4	22.3	209.2	6.5	100.7	9.3	141.6
5	36.9	332.9	4.0	97.5	3.7	158.7
6	38.8	354.5	- 1.1	- 11.3	- 5.7	75.3
7	32.0	215.6	- 4.6	- 278.2	- 14.8	- 161.1
8	23.0	- 146.7	- 1.4	- 757.9	- 18.3	- 605.5

Table 19 - Warping Moment $\overline{M_w}$ for Different Cases of Loading (Unit - Ton Meters^2)

Section	Loading Case I	Loading Case II	Loading Case III	Loading Case IV	Loading Case V	Loading Case VI
0	0	0	0	0	0	0
1	270	- 1336	- 1833	- 2325	- 200	- 171
2	490	- 2145	- 2956	- 3907	- 416	- 331
3	649	- 2112	- 2956	- 4298	- 244	- 43
4	722	- 1178	- 1752	- 3374	396	804
5	683	459	400	- 1341	1255	1998
6	499	2319	1235	1235	1739	2986
7	147	3602	4650	3385	875	2839
8	- 390	3159	4227	3693	- 2726	216

Table 20 - <u>Prestressing Force in Different Sections</u>

<u>of the Beam</u> (Unit - Tons)

Section	N (t)
0	- 9000
1	- 8909
2	- 8818
3	- 8729
4	- 8640
5	- 8552
6	- 8466
7	- 8380
8	- 8295

<u>Note:</u>

-ve sign indicates compression

Coefficient of Cabel

Friction μ = 0.2

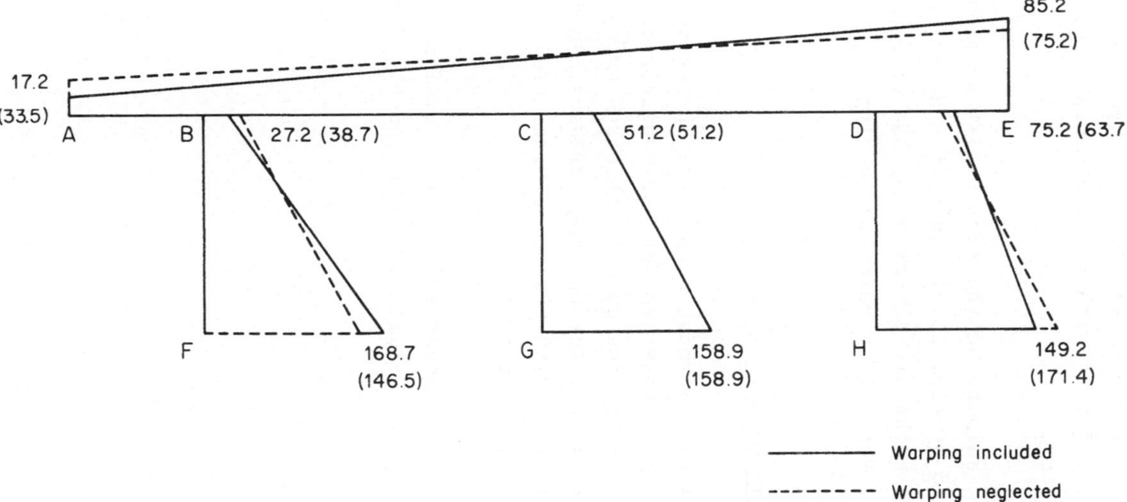

Normal Stress (kg/cm^2) in Concrete for Loading Case 3

FIG. 23

For the given profile of the prestressing cable and for the
given value of the prestressing forces, the normal stresses
in concrete are found to remain compressive in all sections,
and the greatest value is within the assumed permissible
value of 180 kg/cm^2.

The influence of warping on the stress distribution is
clearly indicated by Fig. 23. It shows the distribution of
normal stresses in the cross section of the bridge at the
central support. The values refer to loading case 3.

Discussion of the Results

The results show that warping of a curved prestressed concrete beam with open cross section has considerable influence
on the normal stresses. The warping stresses amounts to between 10 to 20 percent of the bending stresses. An accurate
analysis of such beams with taking warping torsion into
account is therefore essential.

The values of the St. Venant and Warping Torsion Moments in
table 18 show that the Torsion is predominantly Warping Torsion.

6. SUMMARY

The present thesis deals with the general case of bending
and torsion of thin walled beams of unsymmetrical section.
A system of four fourth order ordinary $differential equa-
tions is presented. Closed homogeneous solutions are given
and the particular integrals are obtained for some cases of
loading.

The solutions are used to obtain transport matrices for
straight and curved beams. The theory is extended to beams
of variable cross section using the Finite Elements method.

A short description of the computer program prepared by
the author for the statical analysis of thin walled beams
of open cross section is given in chapter 5.

The problem of skew supports is discussed.

In chapter 5 three illustrative examples based on the present theory are given.

ZUSAMMENFASSUNG

Diese Arbeit befasst sich mit dem allgemeinen Fall der Biegung und Torsion eines dünnwandigen Trägers mit unsymmetrischem Querschnitt. Ein System von vier Differentialgleichungen und für einige Lastfälle werden die partikulären Lösungen bestimmt.

Mit Hilfe dieser Lösungen werden die Elemente der Uebertragungsmatrizen für gerade und gekrümmte Träger gerechnet. Die Methode der endlichen Elemente ermöglicht die Anwendung dieser Theorie für Balken mit variablem Querschnitt.

In Kapitel 4 wird ein vom Verfasser ausgearbeitetes Computer Programm für die statische Berechnung solcher Träger kurz beschrieben.

Drei Zahlenbeispiele, die auf dieser Theorie beruhen, werden in Kapitel 5 gegeben.

RESUME

Ce travail s'occupe du problème de la flexion et torsion
d'une poutre à parois minces avec section asymmétrique.

Un système de 4 équations differentielles avec ses solutions
homogènes et particulières est devévolpé.

A l'aide de ces solutions on détermine les éléments des
matrices de transport pour des poutres droites et courbes.
La méthode des éléments finis rend possible l'application
de cette théorie à des poutres à section variable.

Dans le quatrième chapitre on donne une courte description
du programme pour le calculateur du RZETH (CDC - 1604 A).

On discute aussi le problème des appuis biais.

Le cinquième chapitre présente trois examples se basant sur
cette théorie.

7. NOTATION

7.1 Coordinate System :

O   — The coordinate origin, an arbitrarily chosen point in the cross section.

Ox, Oy — Cartesian axes in the plane of the cross section.

Oz   — Axis perpendicular to the plane of cross section.

Ox, Oy, Oz form a left handed system

7.2 Displacements of the coordinate origin O :

Ux   — positive in the direction of positive x , cm

Uy   — positive in the direction of positive y , cm

Uz   — positive in the direction of positive z , cm

7.3 Rotations

θ — Angle of rotation of the cross section with respect to the chosen axes Ox, Oy, Oz, positive when clockwise as seen from the positive z side, radians

$$\phi x = \frac{dUx}{dz} = \frac{1}{L} \cdot \frac{dUx}{d\zeta} \quad , \quad \text{radians}$$

$$\phi y = \frac{dUy}{dz} = \frac{1}{L} \cdot \frac{dUy}{d\zeta} \quad , \quad \text{radians}$$

ϕx, ϕy positive when the tangent rotates clockwise

$$\text{Unite Twist}: \quad \theta' = \frac{d\theta}{dz} = \frac{1}{L} \cdot \frac{d\theta}{d\zeta} \quad , \quad \text{radians/cm}$$

L   — A reference length of the beam

$$\zeta = \frac{z}{L} \quad , \quad \text{dimensionless}$$

7.4 Material Constants:

E — Modulus of Elasticity, kg/cm^2

G — Shear Modulus, kg/cm^2

7.5 Sectional Properties:

Gx, Gy — Coordinates of the Center of Gravity, cm

Kx, Ky — Coordinates of the Shear Center, cm

$F = \int dF$ — Area, cm^2

$Sy = \int x\,dF$ — Statical Moment, cm^3

$Sx = \int y\,dF$ — Statical Moment, cm^3

$Sw = \int w\,dF$ — Statical Moment (warping), cm^4

$Ix = \int y^2 dF$ — Moment of Inertia w.r.t. x axis, cm^4

$Iy = \int x^2 dF$ — Moment of Inertia w.r.t. y axis, cm^4

$Iw = \int w^2 dF$ — Warping Constant, cm^6

$Ixy = \int xy\,dF$ — Product of Inertia, cm^4

$Iwx = \int wx\,dF$ — Product of Inertia x (warping), cm^5

$Iwy = \int wy\,dF$ — Product of Inertia y (warping), cm^5

$K = \Sigma \frac{1}{3} bt^3$ — St. Venant Torsion Constant, cm^4

7.6 Parameters:

$$a_{11} = 1 - \frac{Sy^2}{FIy} \quad , \text{ dimensionless}$$

$$a_{12} = \frac{Ixy}{Iy} - \frac{Sx\,Sy}{FIy} \quad , \text{ dimensionless}$$

$$a_{13} = \frac{Iwx}{Iy} - \frac{Sw\,Sy}{FIy} \quad , \text{ cm}$$

$$a_{21} = \frac{Ixy}{Ix} - \frac{Sx\,Sy}{FIx} \quad , \text{ dimensionless}$$

$$a_{22} = 1 - \frac{Sx^2}{FIx} \quad , \text{ dimensionless}$$

$$a_{23} = \frac{Iwy}{Ix} - \frac{Sx\,Sw}{FIx} \quad , \text{ cm}$$

$$a_{31} = \frac{Iwx}{Iw} - \frac{Sy\,Sw}{FIw} \quad , \text{ cm}^{-1}$$

$$a_{32} = \frac{Iwy}{Iw} - \frac{Sx\,Sw}{F\,Iw} \quad , \quad cm^{-1}$$

$$a_{33} = 1 - \frac{Sw^2}{F\,Iw} \quad , \quad \text{dimensionless}$$

$$a_{34} = \frac{G\,K}{E\,Iw}\,L^2 \quad , \quad \text{dimensionless}$$

$$b_{11} = a_{11} - \frac{a_{12}\,a_{21}}{a_{22}} \quad , \quad \text{dimensionless}$$

$$b_{12} = a_{13} - \frac{a_{12}\,a_{23}}{a_{22}} \quad , \quad cm$$

$$b_{14} = -\frac{a_{12}}{a_{22}} \quad , \quad \text{dimensionless}$$

$$b_{21} = a_{31} - \frac{a_{32}\,a_{21}}{a_{22}} \quad , \quad \text{dimensionless}$$

$$b_{22} = a_{33} - \frac{a_{23}\,a_{32}}{a_{22}} \quad , \quad \text{dimensionless}$$

$$b_{23} = a_{34} = \frac{G\,K}{E\,Iw}\,L^2 \quad , \quad \text{dimensionless}$$

$$b_{24} = -\frac{a_{32}}{a_{22}} \quad , \quad cm^{-1}$$

$$c_{11} = b_{22} - \frac{b_{21}\,b_{12}}{b_{11}} \quad , \quad \text{dimensionless}$$

$$c_{12} = b_{23} = \frac{G\,K}{E\,Iw}\,L^2 \quad , \quad \text{dimensionless}$$

$$X_1 = \frac{L^3}{E\,F} \quad , \quad kg^{-1}\,cm^3$$

$$X_2 = \frac{L^4}{E\,Iy} \quad , \quad kg^{-1}\,cm^2$$

$$X_3 = \frac{L^4}{E\,Ix} \quad , \quad kg^{-1}\,cm^2$$

$$X_4 = \frac{L^4}{E\,Iw} \quad , \quad kg^{-1}$$

$$\alpha = -\frac{b_{12}}{b_{11}} \quad , \quad cm$$

$$\beta = \sqrt{\frac{c_{12}}{c_{11}}} \quad , \quad \text{dimensionless}$$

$$\gamma = -\frac{b_{21}}{b_{11}} \quad , \quad cm^{-1}$$

$$A = \frac{X_4}{c_{11}} \quad , \quad kg^{-1}$$

$$B = \frac{\gamma \cdot X_2}{c_{11}} \quad , \quad kg^{-1}\,cm$$

$$C = \frac{g_{17} \cdot X_3}{c_{11}} \quad , \quad \mathrm{kg^{-1}\ cm}$$

$$D = \frac{g_{18} \cdot X_1}{c_{11}} \quad , \quad \mathrm{kg^{-1}\ cm}$$

$$A_0 = - \frac{A}{2\beta^2} \quad , \quad \mathrm{kg^{-1}}$$

$$B_0 = - \frac{B}{2\beta^2} \quad , \quad \mathrm{kg^{-1}\ cm}$$

$$C_0 = - \frac{C}{2\beta^2} \quad , \quad \mathrm{kg^{-1}\ cm}$$

$$D_0 = - \frac{D}{2\beta^2} \quad , \quad \mathrm{kg^{-1}\ cm}$$

$$A_1 = \alpha\, A_0 \quad , \quad \mathrm{kg^{-1}\ cm}$$

$$B_1 = \alpha\, B_0 \quad , \quad \mathrm{kg^{-1}\ cm^2}$$

$$C_1 = \alpha\, C_0 \quad , \quad \mathrm{kg^{-1}\ cm^2}$$

$$D_1 = \alpha\, D_0 \quad , \quad \mathrm{kg^{-1}\ cm^2}$$

$$g_1 = - \frac{a_{23}}{a_{22}} \quad , \quad \mathrm{cm}$$

$$g_2 = - \frac{a_{21}}{a_{22}} \quad , \quad \text{dimensionless}$$

$$g_3 = \frac{S\,w}{F} \quad , \quad \mathrm{cm^2}$$

$$g_4 = \frac{S\,y}{F} \quad , \quad \mathrm{cm}$$

$$g_5 = \frac{S\,x}{F} \quad , \quad \mathrm{cm}$$

$$g_6 = \frac{S\,x}{I\,x} \quad , \quad \mathrm{cm^{-1}}$$

$$g_7 = \frac{I_{xy}}{I\,x} \quad , \quad \text{dimensionless}$$

$$g_8 = \frac{I_{wy}}{I\,x} \quad , \quad \mathrm{cm}$$

$$g_9 = \frac{S\,y}{I\,y} \quad , \quad \mathrm{cm^{-1}}$$

$$g_{10} = \frac{I_{xy}}{I\,y} \quad , \quad \text{dimensionless}$$

$$g_{11} = \frac{I_{wx}}{I\,y} \quad , \quad \mathrm{cm}$$

$$g_{12} = \frac{S\,w}{I\,w} \qquad , \ \mathrm{cm}^{-2}$$

$$g_{13} = \frac{I\,wx}{I\,w} \qquad , \ \mathrm{cm}^{-1}$$

$$g_{14} = \frac{I\,wy}{I\,w} \qquad , \ \mathrm{cm}^{-1}$$

$$g_{15} = g_9 + b_{14} \cdot g_6 \qquad , \ \mathrm{cm}^{-1}$$

$$g_{16} = g_{12} + b_{24} \cdot g_6 \qquad , \ \mathrm{cm}^{-2}$$

$$g_{17} = b_{24} + \gamma \cdot b_{14} \qquad , \ \mathrm{cm}^{-1}$$

$$g_{18} = g_{16} + \gamma \cdot g_{15} \qquad , \ \mathrm{cm}^{-2}$$

$$\alpha_2 = \frac{x_2}{24 \cdot b_{11}} \qquad , \ \mathrm{kg}^{-1}\,\mathrm{cm}^2$$

$$\alpha_3 = \frac{x_3 \cdot b_{14}}{24 \cdot b_{11}} \qquad , \ \mathrm{kg}^{-1}\,\mathrm{cm}^2$$

$$\alpha_4 = \frac{x_1 \cdot g_{15}}{24 \cdot b_{11}} \qquad , \ \mathrm{kg}^{-1}\,\mathrm{cm}^2$$

$$\gamma_1 = \frac{x_3}{24 \cdot a_{22}} \qquad , \ \mathrm{kg}^{-1}\,\mathrm{cm}^2$$

$$\gamma_2 = \frac{x_1 \cdot g_6}{24 \cdot a_{22}} \qquad , \ \mathrm{kg}^{-1}\,\mathrm{cm}^2$$

$$\gamma_3 = \frac{x_1}{6} \qquad , \ \mathrm{kg}^{-1}\,\mathrm{cm}^3$$

$$m_1 = g_1 + \alpha\, g_2 \qquad , \ \mathrm{cm}$$

$$m_2 = g_3 + \alpha\, g_4 + m_1 g_5 \qquad , \ \mathrm{cm}^2$$

$$m_3 = g_4 + g_2 g_5 \qquad , \ \mathrm{cm}$$

$$\phi_1 = \frac{G\,K}{E\,I\,w}\,L^2 \qquad , \ \text{dimensionless}$$

$$P_1 = \beta^2(g_{12}m_2 - \alpha\, g_{13} - m_1 g_{14} - 1) + \phi_1 \beta \qquad , \ \text{dimensionless}$$

$$P_2 = 2\,(m_3 g_{12} - g_{13} - g_2 g_{14}) \qquad , \ \mathrm{cm}^{-1}$$

$$P_3 = 2\,(g_5 g_{12} - g_{14}) \qquad , \ \mathrm{cm}^{-1}$$

$$P_4 = \beta^2(g_{12}m_2 - \alpha\, g_{13} - g_{14} m_1 - 1) \qquad , \ \text{dimensionless}$$

$$P_5 = \beta^2(g_9 m_2 - \alpha - g_{10} m_1 - g_{12}) \quad , \text{ cm}$$

$$P_6 = 2\,(m_3 g_9 - 1 - g_2 g_{10}) \quad , \text{ dimensionless}$$

$$P_7 = 2\,(g_9 g_5 - g_{10}) \quad , \text{ dimensionless}$$

$$P_8 = \beta^2(g_6 m_2 - \alpha\, g_7 - m_1 - g_8) \quad , \text{ cm}$$

$$P_9 = 2\,(m_3 g_6 - g_7 - g_2) \quad , \text{ dimensionless}$$

$$P_{10} = 2\,(g_6 g_5 - 1) \quad , \text{ dimensionless}$$

$$P_{11} = \beta\, P_5 \quad , \text{ cm}$$

$$P_{12} = \beta\, P_8 \quad , \text{ cm}$$

$$E_1 = L\,Uz_p{}' - g_4 Ux_p{}'' - g_5 Uy_p{}'' - g_3\, \theta'' \quad , \text{ cm}^2$$

$$E_2 = g_{12} L\,Uz_p{}'' - g_{13} Ux_p{}''' - g_{14} Uy_p{}''' - \theta_p{}''' + \phi_1 \theta_p{}' \quad , \text{ dimensionless}$$

$$E_3 = g_{12} L\,Uz_p{}' - g_{13} Ux_p{}'' - g_{14} Uy_p{}'' - \theta_p{}'' \quad , \text{ dimensionless}$$

$$E_4 = g_9\, L\,Uz_p{}' - Ux_p{}'' - g_{10} Uy_p{}'' - g_{11} \theta_p{}'' , \text{ cm}$$

$$E_5 = g_6\, L\,Uz_p{}' - g_7\, Ux_p{}'' - Uy_p{}'' - g_8 \theta_p{}'' , \text{ cm}$$

$$E_6 = g_9\, L\,Uz_p{}'' - Ux_p{}''' - g_{10} Uy_p{}''' - g_{11} \theta_p{}''' , \text{ cm}$$

$$E_7 = g_6\, L\,Uz_p{}'' - g_7\, Ux_p{}''' - Uy_p{}''' - g_8 \theta_p{}''' , \text{ cm}$$

Here p indicates particular integral

8. BIBLIOGRAPHY

1)   Vlassov V.Z. - "Theory of thin walled elastic beams", translated from Russian, published by the Israel Programme of Scientific Translations, Jerusalem.

2)   Cywinski Z. - "Torsion des dünnwandigen Stabes mit veränderlichem, einfach symmetrischem, offenem Querschnitt", "Der Stahlbau", Oktober 1964.

3)   Becker G. - "Ein Beitrag zur statischen Berechnung beliebig gelagerter ebener gekrümmter Stäbe mit einfach symmetrischen dünnwandigen offenen Profilen von in der Stabachse veränderlichem Querschnitt unter Berücksichtigung der Wölbtorsion". "Der Stahlbau", November/ Dezember 1965.

4)   Bazant Z.P. - "Non-uniform torsion of thin walled beams of variable section". Publication of the International Association for Bridge and Structural Engineering, 1965, Zurich.

5)   Heilig R. - "Beitrag zur Theorie der Kastenträger beliebiger Querschnittsform". "Der Stahlbau", November 1961.

6)   Karamuk E. - "Zur Berechnung dünnwandiger Stäbe mit variablem offenem Querschnitt", Dissertation, ETH Zürich, 1968.

7)   Zurmühl - "Matrizen und ihre technischen Anwendungen", Springer Verlag, Berlin.

8)   Wilde P. - "The torsion of thin-walled bars with variable cross section", Archiwum Mechaniki Stosowanej, 4,20 (1968).

APPENDIX

$$\text{STATE VECTOR} = \text{MATRIX } X \times \text{VECTOR CON}$$

STATE VECTOR	1	2	3	4	5	6	7	8	9	10	11	12	13	14	(particular)	VECTOR CON
θ	1	ζ	$\sinh\beta\zeta$	$\cosh\beta\zeta$	0	0	0	0	0	0	0	0	0	0	θp	C_1
θ'	0	1	$\beta\cosh\beta\zeta$	$\beta\sinh\beta\zeta$	0	0	0	0	0	0	0	0	0	0	$\theta p'$	C_2
U_x	a	$a\zeta$	$a\sinh\beta\zeta$	$a\cosh\beta\zeta$	1	ζ	ζ^2	ζ^3	0	0	0	0	0	0	$U_x p$	C_3
U_x'	0	a	$a\beta\cosh\beta\zeta$	$a\beta\sinh\beta\zeta$	0	1	2ζ	$3\zeta^2$	0	0	0	0	0	0	$U_x p'$	C_4
U_y	m_1	$m_1\zeta$	$m_1\sinh\beta\zeta$	$m_1\cosh\beta\zeta$	G_2	$G_2\zeta$	$G_2\zeta^2$	$G_2\zeta^3$	1	ζ	ζ^2	ζ^3	0	0	$U_y p$	C_5
U_y'	0	m_1	$m_1\beta\cosh\beta\zeta$	$m_1\beta\sinh\beta\zeta$	0	G_2	$2G_2\zeta$	$3G_2\zeta$	0	1	2ζ	$3\zeta^2$	0	0	$U_y p'$	C_6
$L\cdot U_z$	0	m_2	$m_2\beta\cosh\beta\zeta$	$m_2\beta\sinh\beta\zeta$	0	m_3	$2m_3\zeta$	$3m_3\zeta^2$	0	G_5	$2G_5\zeta$	$3G_5\zeta^2$	1	ζ	$L\cdot U_z p$	C_7
$N\cdot\dfrac{L^2}{EF}$	0	0	0	0	0	0	0	0	0	0	0	0	0	1	E_1	C_8
$T\cdot\dfrac{L^3}{EIw}$	0	ϕ_1	$P_1\cosh\beta\zeta$	$P_1\sinh\beta\zeta$	0	0	0	$3P_2$	0	0	0	$3P_3$	0	0	E_2	C_9
$Mw\cdot\dfrac{L^2}{EIw}$	0	0	$P_4\sinh\beta\zeta$	$P_4\cosh\beta\zeta$	0	0	P_2	$3P_2\zeta$	0	0	P_3	$3P_3\zeta$	0	G_{12}	E_3	C_{10}
$-My\cdot\dfrac{L^2}{EIy}$	0	0	$P_5\sinh\beta\zeta$	$P_5\cosh\beta\zeta$	0	0	P_6	$3P_6\zeta$	0	0	P_7	$3P_7\zeta$	0	G_9	E_4	C_{11}
$Mx\cdot\dfrac{L^2}{EIx}$	0	0	$P_8\sinh\beta\zeta$	$P_8\cosh\beta\zeta$	0	0	P_9	$3P_9\zeta$	0	0	P_{10}	$3P_{10}\zeta$	0	G_6	E_5	C_{12}
$Qx\cdot\dfrac{L^3}{EIy}$	0	0	$P_{11}\cosh\beta\zeta$	$P_{11}\sinh\beta\zeta$	0	0	0	$3P_6$	0	0	0	$3P_7$	0	0	E_6	C_{13}
$Qy\cdot\dfrac{L^3}{EIx}$	0	0	$P_{12}\cosh\beta\zeta$	$P_{12}\sinh\beta\zeta$	0	0	0	$3P_9$	0	0	0	$3P_{10}$	0	0	E_7	C_{14}
1	0	0	0	0	0	0	0	0	0	0	0	0	0	0	1	1